William Hampson

Radium Explained

A Popular Account of
the Relations of Radium to
the Natural World, to Scientific Thought
and to Human Life

Elibron Classics
www.elibron.com

Elibron Classics series.

ISBN 0-543-69175-6 (paperback)
ISBN 0-543-69174-8 (hardcover)

This Elibron Classics Replica Edition is an unabridged facsimile of the edition published in 1905 by T. C. & E. C. Jack, London.

RADIUM EXPLAINED

RADIUM EXPLAINED

A POPULAR ACCOUNT OF THE RELATIONS
OF RADIUM TO THE NATURAL WORLD,
TO SCIENTIFIC THOUGHT, AND
TO HUMAN LIFE

BY

Dr W. HAMPSON, M.A.

FORMERLY OPEN SCHOLAR OF TRINITY COLLEGE, OXFORD; LECTURER UNDER THE
LONDON COUNTY COUNCIL AT UNIVERSITY COLLEGE, LONDON, 1900-1901;
AUTHOR OF PAPERS FOR THE SOCIETY OF CHEMICAL INDUSTRY,
NATURE, AND THE TECHNICAL JOURNALS; INVENTOR
OF THE HAMPSON AIR-LIQUEFIER

WITH ILLUSTRATIVE DIAGRAMS

LONDON: T. C. & E. C. JACK
34 HENRIETTA STREET, W.C.
AND EDINBURGH
1905

TABLE OF CONTENTS.

LIST OF ILLUSTRATIONS.

PREFACE.

Notice is hereby given that the following pages are not intended to serve as a text-book filled with facts and demonstrations arranged in the most convenient way for committing to memory with a view to a subsequent examination. Those who wish for a full account of the principles of radio-activity suitable for advanced students of physics, cannot do better than read the admirable works of Professors Rutherford and Thomson and Mr Soddy. But those works, like the papers on the subject communicated to the Royal Society, either bristle with mathematical formulæ, sometimes for pages together, or employ so many technical scientific phrases and implications as to be quite unintelligible to the ordinary lay reader.

Radium, however, and its lessons are matters of great importance and interest to all; and they deserve to be set forth in such a way that they may be understanded of the people. This little book is written, accordingly, for the people—the more thoughtful section of course—that they may obtain some more systematic and intelligible information about radium than could be gathered from frequent disconnected and sensational articles in the daily press, or smart and "snappy" contributions to the monthly magazines.

This account is intended to make clear, in the first place, what there is in the processes underlying the nature and actions of radium to give them such a unique hold on the consideration of the scientific world, and to justify so much general interest and surprise. Secondly, to show how the knowledge obtained in connection with the study of radium has modified scientific views as to the constitution of the material universe in which we live. Thirdly, to set forth impartially the limits of the practical benefit which we get from radium now, and the hopes that may be rationally entertained in that direction for the future.

This is a work of explanation. The scientific student must have the proofs of a proposition set forth in detail, that he may follow and justify them step by step. For the general reader it is more useful to explain the bearings of the proposition, and to explain the nature of the proof, in words and ideas which are not so unfamiliar as to deter him from the effort to understand. This object the author has kept so strictly in view, that he has not permitted himself to introduce so much as one equation; and if one single expression of an apparently algebraic nature has been inevitably employed, it has been so simplified as to be algebraic only in form. The whole book can be understood without more mathematical, physical, or chemical knowledge than form part of the common intellectual currency of mankind.

RADIUM EXPLAINED.

I.

The Puzzle of Radium.

For more than twelve months radium has received an amount of public attention which is not often bestowed on a strictly scientific subject. Everyone is now familiar with the word at least, which has obtained such wide-spread recognition, that, besides seeing radium dances in ballets, we can buy radium collars, radium stoves, and radium polish. This sudden and extensive vogue is the more remarkable, since radium has not been made the vehicle of so many extravagant and dishonest promises of immediate revolutions in the industrial world, as was the case with electricity and liquid air in the early days of their development. The fame of radium has grown almost entirely from its really interesting and astonishing position in the scientific world. Its remarkable activities are shown on so small a scale in the minute specimens of it which alone exist, that one would hardly expect them to rouse the interest of many who can be easily impressed by the mighty power of the engines in an Atlantic liner, or the thousands of miles traversed by wireless messages passing from England to America. The fact that the general public have been so widely interested in radium, and so deeply impressed by it, is a remarkable testimony to the high position held at present by science, since the public have had to rely, for the most part, on their faith in the teachings of scientific men, both for their knowledge of the things that radium can do, and for their belief that its doings are surprising and deserving of the most careful attention.

What, then, can radium do, and why are its doings of a nature to cause such surprise and interest among men of science? We will answer the latter question first. Radium, then, in the early days of our acquaintance with it, appeared to afford a contradiction to two of the greatest and most firmly established laws of nature—the conservation of energy and the persistence of matter. It is one of the greatest generalisations of modern science, that energy, however it may be changed from one form into another, is never, in our experience, destroyed on the one hand, or originated on the other. Whenever we see energy displayed, we can always trace it, if we have sufficient knowledge of the facts, to some previously existing form or forms of energy. The movement of the looms in a factory is a transformation of the expansion energy of the steam, that is, the movement of the steam molecules striking one another and driving one another farther apart. This is a transformation of the heat energy of the gases of combustion. The intra-atomic energies, which were transformed into a combustion-rush, were themselves transformations of the actinic vibrations of the ether which acted on the leaves of the growing plants out of which the coal was formed. These actinic vibrations were set up, at a distance of many millions of miles, by the vibrating particles of the sun: and if we are not yet all agreed as to exactly whence and how the sun obtained and maintains his enormous heat energy (radium is now one of the alleged sources), we are at any rate quite convinced that when a theory on this subject is generally accepted, it will prove to be one more instance of the great law that, wherever energy appears, it has been transformed from some previous manifestation of energy. If we take it the other way about, beginning with the sun, and ending with the factory looms, we readily perceive the converse truth, that, wherever energy disappears, it is not really destroyed, but only transformed into some other kind of movement, or movement in some other substance. Moreover, if we examine the matter quantitatively, weighing and measuring the energies involved in each successive pair of manifestations, we find that the force developed is always exactly equal to that out of which it was transformed. If the travelling energy of a moving train be added to the friction-heat energy of the wheels, axles, rails, and disturbed air, to the heat energy passing through the sides of the boiler into the atmosphere, and to the heat energy imparted to the steam, the gases of combustion, and the ashes, it is found that the sum total of energy is exactly equal to the heat energy developed by the burning fuel. From the complicated steam engine to the simple lever, it is found, on careful measurement, to be true of every kind of machine, that the work got out of it is exactly equal to the work put into it. There was a time, and not very long ago, when this truth was not generally recognised. People thought they could work magical tricks with energy, transforming a smaller quantity of work into a greater by means of cunning mechanical devices. Every perpetual motion machine in the Patent Office is pathetic evidence of the once prevalent belief that less energy could be transformed into more without the assistance of a miracle. The opposite truth may be held to have been finally established when Joule made his great demonstration of the mechanical equivalent of heat, showing that a definite amount of mechanical work is always equal to a definite amount of heat, and can be converted into that amount, neither more nor less. The same truth applies to the transformation of all other manifestations of energy, whether in the form of light, electricity, chemical action, or activities of nerve and muscle. This is the great law or truth of the conservation of energy,—that, within our experience, energy is never created or destroyed, and that, when it appears or disappears, it is always transformed into a quantity or quantities of energy, in some other places or forms but exactly equal in amount to the original energy.

This great law, then, radium appears to disobey. In the first place, radium is constantly throwing off heat, without itself getting any colder, although it has no apparent source of supply from which it can replace the heat it has lost; thus laying itself open to the same suspicions as those who are seen to be constantly spending money without having any visible means of subsistence. Curie and Laborde found that less than 3 grains of radium, which was present in five times its weight of barium chloride, kept the mixture nearly 3° Fahrenheit warmer than surrounding objects. Now we know that a foot-warmer will gradually cool down to the temperature of surrounding objects. The mere fact that we feel an object to be warm, implies that it is giving up warmth or heat to us, that is to say, it is getting cooler. Heat, like sound, is a form of energy, that is, of movement of matter, and, like other forms of energy, becomes exhausted in the material at its source by dissipation through surrounding material. It may, however, be replaced as fast as it is dissipated.

Thus one end of a poker, which is constantly giving heat energy to the surrounding ether, may be constantly receiving fresh heat energy through the other end from a hot fire; and while that is the case, the outer end need not grow cooler. But with radium it is different. This wonderful substance can keep warm, can therefore keep giving heat energy to the ether and other things around, without itself growing any cooler, although the dissipated energy is not replaced from any visible external source, such as a fire, an electric apparatus, friction, or hammering. It is as if a foot-warmer could keep hot indefinitely without having the water changed, or being placed on a stove, or getting a renewal, from any external source, of the heat energy which it is constantly sending out in all directions. Curie found later, that a larger quantity of radium, nearly 11 grains, kept at a temperature of 6° F. above that of its surroundings; and Giesel, examining 15½ grains of radium, proved that it maintained an excess temperature of 9° F.

A very striking experiment to demonstrate this constant and undiminishing output of heat energy by radium, is the boiling of liquids, as shown by Curie at the Royal Institution in 1903. If a liquid be kept just at its boiling-point, but not quite boiling, a very little additional heat will not make the liquid any hotter, but will cause the sudden conversion of some of the liquid into vapour, with the formation of bubbles, which constitutes boiling. Therefore, if a substance like radium, which is constantly giving off heat, be placed in a liquid maintained in that condition, at the boiling-point, but not actually boiling, the additional heat supplied locally by the presence of the radium will cause a local boiling of the liquid in its neighbourhood, and so make clearly visible the fact that heat is being given off by the object inserted. It is not easy to keep a liquid just at the boiling-point without actually boiling. The difficulty is least if we choose a liquid whose boiling-point is below the ordinary temperature, such as liquid air, which was employed by Curie. A quantity of liquid air, placed in an ordinary vessel, is kept boiling by the heat of the surrounding air so vigorously as to mask the small amount of boiling produced by inserting in it some small object, such as a piece of radium. To prevent this, the liquid air is kept in a vacuum vessel, of Dewar's improved form, shown in Fig. 1, A. This consists of two test-tubes, one inside the other, and welded together at the top, so that an enclosed air-space surrounds the smaller one. The air is then pumped out of this space through a small opening which is afterwards welded over, leaving the inner tube surrounded by a vacuum space, which contains no air or other substance to conduct external heat to any object in the inner tube. Still, external objects can radiate some heat across the vacuum space through the ether, which carries heat vibrations as air carries sound; and this small amount of heat causes a little boiling of the liquid air, just enough to spoil the experiment. To obviate this, the vessel containing the liquid air is surrounded by liquid air contained in a similar but larger vessel, as at B. All heat radiated from outside is now caught by the outer layer of liquid air, part of which is boiled away, protecting thereby the liquid air in the inner vessel, which lies perfectly still and free from bubbles. If now a piece of glass tube or any other object at ordinary temperature be dropped into the liquid air in the centre, the liquid begins to boil rapidly under the influence of the heat of this object, which, in comparison with its intense cold, may be said to be intensely hot. The boiling, however, ceases before long, because the object inserted has cooled down to the temperature of the surrounding liquid; and all is as still as before. Now drop into the liquid air a small glass tube containing some radium—a gramme if possible—but less will serve. This also at first causes rapid boiling of the liquid air, in comparison with which both the tube and the radium are intensely hot. But presently the tube is cooled down nearly to the same temperature as the liquid air, and the rapid boiling of the latter ceases. In this case,

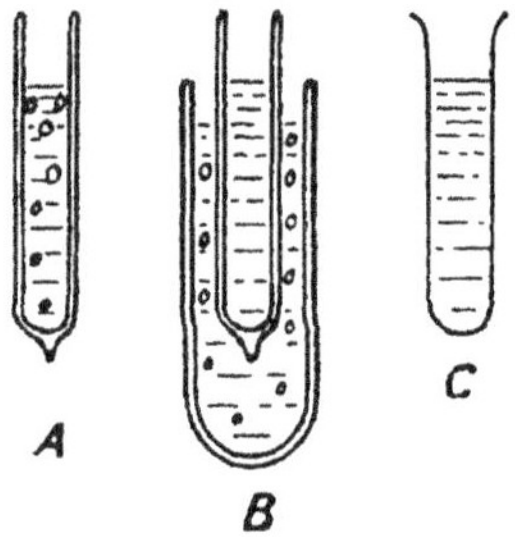

Fig. 1.

however, a slow boiling continues when the rapid boiling is over; the radium, apparently, does not cool down quite to the temperature of the liquid air. Now, for a comparison experiment, pour boiling water into an ordinary large test-tube, as at C. The moment it is poured out it ceases to boil, having fallen very slightly below the boiling-point; but if a piece of red-hot iron, such as a large nail, be dropped into it, this will produce brisk boiling of the water. This boiling, again, ceases shortly, and ceases altogether, while the liquid air, with the radium in it, is still slowly boiling. And if the liquid air in the inner and outer vessels be kept replenished, the boiling caused by the radium goes on for days, months, and years. This looks like a miracle. The marvel becomes greater if we carefully examine afterwards the radium which has acted in this wonderful way. After parting with heat in this way for a year, it is no cooler than it was at the end of the first hour. Weigh it carefully before and after the year's work. The coals that boil water on a fire weigh much less after an hour's work, and in a few hours are almost entirely burnt away. But the most delicate balance, weighing to the thousandth part of a grain, cannot detect the slightest loss of weight in the radium which has been hard at work for a year boiling liquid air. Lastly, no chemical or physical change can be detected to account for the output of heat energy. When heat energy is produced by the union of quicklime with water, there is a corresponding change in the constitution of these materials: they have lost latent or potential energy either in the form of diminished chemical activity or diminished extension, or both. But in the case of radium, it is not possible, even after it has been working a whole year, to detect the smallest degradation change in either its chemical or its physical properties. Nothing observable has happened to it except a slight darkening of the colour of its crystals. Here, then, we appear to be in presence of an exception to that great and well established natural truth, the doctrine of the conservation of energy. All other machines or engines for giving out work must be constantly supplied with energy in some form, such as the falling of water, the blowing of wind, the vibrations within electric wires, or the combustion of fuel. When the supply of energy fails, the machine ceases to work.

A very interesting and beautiful little scientific toy, depending on the light exciting quality of radium, has been invented by Sir William Crookes It is called the spinthariscope, which means spark-viewer. Fig. 2, a sectional sketch of this instrument, shows that it consists of a small chamber, *a*, beneath a lens of low power, *b*. Near the bottom of the chamber is a small holder, *c*, projecting from the side, and having its point covered with a very minute quantity of radium The floor of the chamber, *d*, is covered with a screen of zinc-blende, which is made fluorescent to the eye by the action of the small quantity of radium above it. The lens, however, when the screen is viewed through it, analyses this fluorescence into a multitude of small sparks, each of which lasts only for a small fraction of a second. On the outer part of the screen these are seen individually as separate sparks; but towards the centre, where they are more numerous, they coalesce into waves and splashes of light. The accepted explanation of this display will be given later; meantime, it may be remarked that it is a most wonderful and beautiful instance of the surprising power of radium to keep up an unflagging activity, apparently for ever.

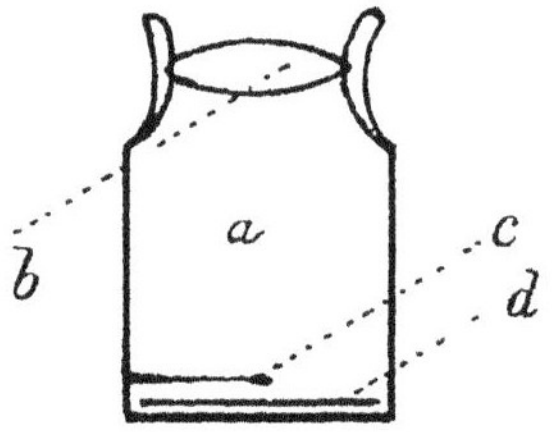

Fig. 2.

A third form in which radium is constantly giving out energy is that of certain rays, which, like the Rontgen rays, can pass through opaque materials and then act upon a photographic plate. As with the Rontgen rays, the difficulty they have in passing through various materials is proportionate to the density of the latter, so that while they pass with almost undiminished intensity through black paper, wood, or thin aluminium, they are much weakened by passing through dense metals. If, then, when the radium rays are acting upon a plate, some such objects be interposed, the portions covered by them will be less strongly acted on than the rest of the plate, and so an image of the shape, though not the surface, of the object will be produced, much resembling a radiogram made with Röntgen rays. Fig. 3 shows a plate acted on directly by radium and uranium in pitch-blende.

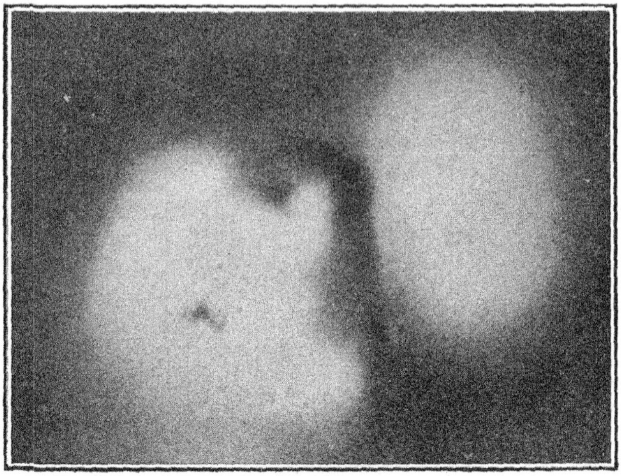

Fig. 3.

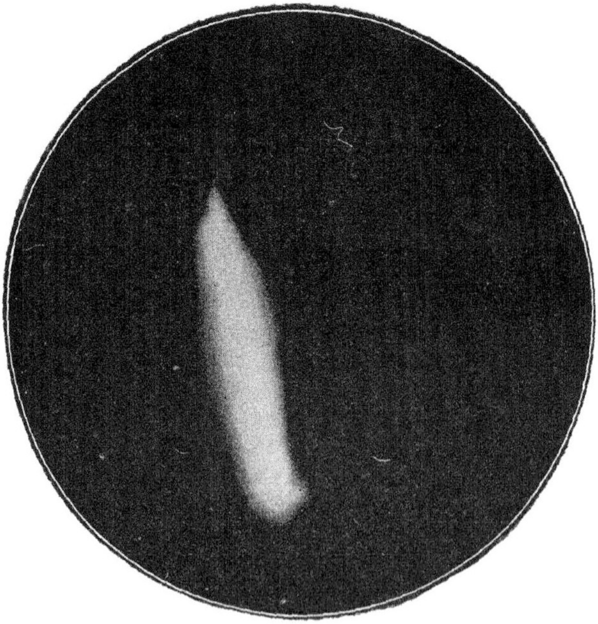

Fig. 4.

Fig. 4 shows how a more concentrated specimen, though containing a large excess of barium as an impurity, was able to act upon a plate through the glass of the tube which contained the specimen, through the side of a wooden box in which it was also placed, and through dense black paper with which the plate was covered. In Fig. 5 we see Soddy's illustration of the extremely energetic action of a very pure sample of radium, which was enclosed in a small glass tube. This tube was held like a pencil, and the end containing the

Fig. 5.

small speck of radium moved, as in writing, just over the plate, with densely opaque black paper intervening. Rapidly as it moved, the radium acted on the plate through the glass and the black paper so energetically as to leave the impression of its action, in written letters, on the plate.

Fig. 6 shows a skiagram of match-box and gold and silver coins taken through black and yellow paper. Fig. 7 shows a coin, key, and heart-shaped key-ring taken in the same way. The images in Fig. 8, sleeve-link, watch-key, button, and small padlock, were formed by rays passing through black and yellow paper and the sheet-iron and tin of a tin-box lid from a specimen of radium placed 8 inches above them. It is interesting as showing the strong effect of some tiny specks of radium at close quarters, though still through the same screens. The specks were some loose bits that fell from the radium above on to the lid, which was immediately over the objects, and it

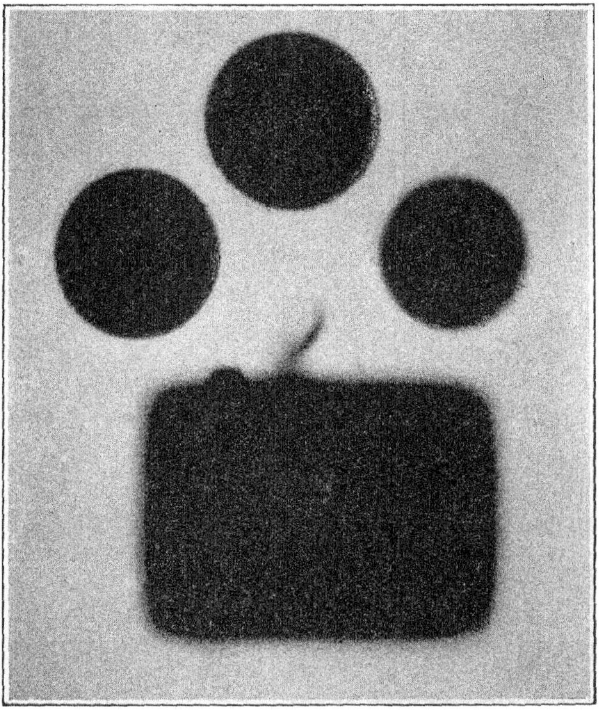

Fig. 6.

can be seen how most of them grouped themselves along the circular groove which surrounded the lid.

In the fourth place, radium is constantly producing electricity with which it charges itself, and a special arrangement can make this constant action manifest. Fig. 9 shows Strutt's apparatus for this purpose. This is a glass vessel

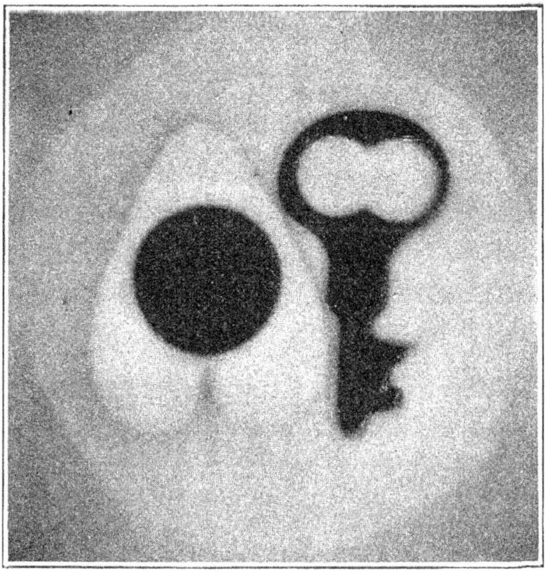

Fig. 7.

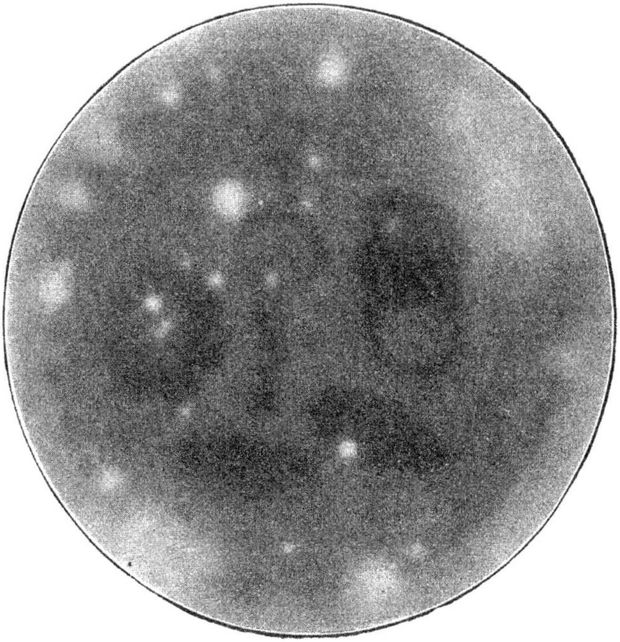

Fig. 8.

from which all the air is pumped out, so as to make a vacuum as perfect as possible. Inside is suspended a small glass tube, *a*, which contains a little radium. This tube is made a conductor of electricity by applying phosphoric acid, and it has attached to its lower end a small electroscope, the leaves of which are shown at *bb*. Strips of foil, *cc*, connect the inside of the vessel to earth. The radium, in a manner to be described later, is constantly at work accumulating in itself a positive charge of electricity, which repels to the leaves an induced positive charge; and the leaves, both having now a positive charge, repel each other farther and farther as the charge accumulates. In a few minutes the charge is sufficient to make the leaves repel each other so far that their extremities touch the earth connection at the side of the vessel, and they are immediately discharged and collapse. The charge then begins to accumulate again, repeating once more the separation of the leaves, which are ultimately discharged again on touching the sides. This work is kept up, and will apparently be kept up for ever, without any observable loss or change of the radium inside, and without any entrance of auxiliary energy from outside.

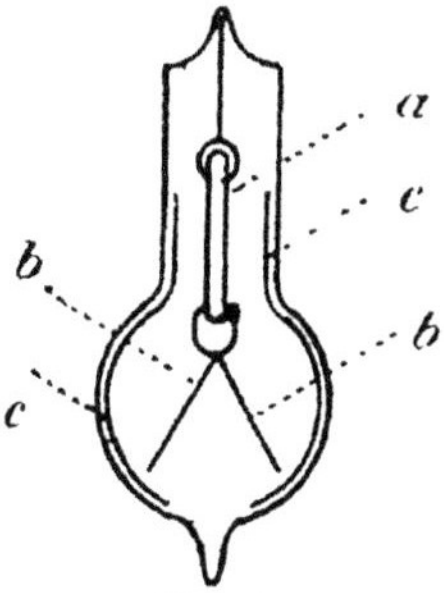

Fig. 9.

Another form of work that radium is constantly doing is the ionization of the surrounding air. Dry air is not a conductor of electricity. But if the molecules and atoms of nitrogen and oxygen which compose the air are broken up into smaller parts called ions, each of which is capable of carrying an electric charge, the air is in that way made into a conductor. The Rontgen rays have the power of acting upon air in this way. It appears that radium can do the same thing for the air in its neighbourhood. Fig. 10 represents an ordinary gold-leaf electroscope, which is seen

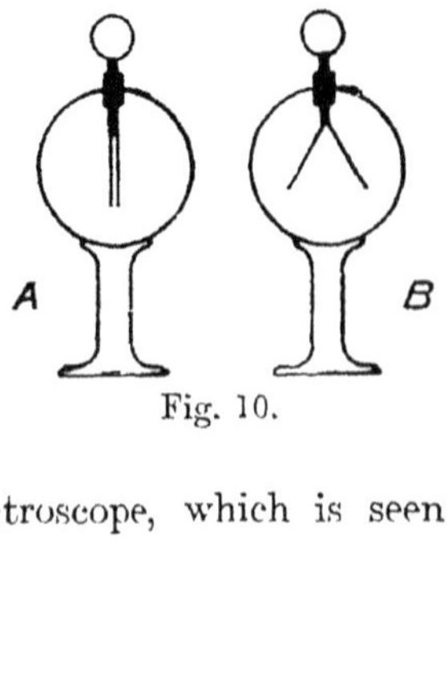

Fig. 10.

in a quiescent state at A. If a piece of ebonite, electrically charged by rubbing, be brought into contact with the knob, the leaves become charged; and as they both receive similar charges, they repel each other and hang apart at the lower end. If the knob be now connected directly or indirectly with earth, this excess charge will be conducted away, and the leaves fall together again by their own weight, slowly if the conducting agency is feeble, quickly if it is strong. When the air is fairly dry, it is such a poor conductor that the charge remains in the electroscope a long time, and the leaves continue parted. If, however, a piece of radium is brought within range of the knob, the leaves at once begin to fall together, showing that the air around the instrument is now able to conduct electricity. The radium is

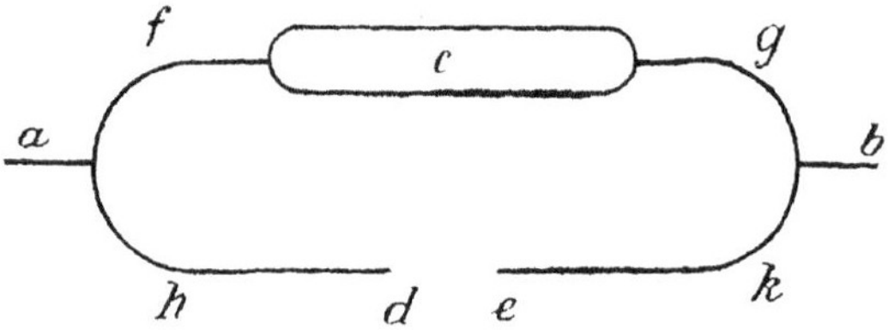

Fig. 11.

acting upon it—ionizing it. This action can be shown effectively in another way. A current from a battery is offered at one part of its circuit two alternative courses, as shown in Fig. 11. If it pass by *fg*, it has to light up on its way a Crookes' tube or an X-ray tube, as at *c*. If it pass by *hk*, it has to leap across a spark-gap, as at *de*. This gap is arranged of such a width that the difficulty of crossing it is a very little greater than that of traversing and illuminating the tube. Accordingly the latter course is that which the current takes, and the tube is kept lit up. If, however, a little radium be brought within range of the spark-gap, it ionizes the air in the gap, making it a conductor of electricity, and so making it easier for the electric current to cross the gap than to pass through the tube. The tube, therefore, is immediately darkened as the current flows by *hk*. When the radium is removed, the gap *de* again becomes too difficult to pass, and the tube is lit up again.

Other phenomena of a chemical or physical kind show that radium is constantly putting forth active forces capable of influencing other objects. A radium salt, if dissolved in water, decomposes it. In the course of a day, hydrogen and oxygen gases are produced in this way to the extent of more than one hundred times the volume of the radium itself. It can change substances into allotropic forms, converting oxygen gas into the denser form of the element known as ozone, and yellow phosphorus into red. It can cause the oxidation of mercury and carbon at ordinary temperatures. And it can produce a violet or brown colour in the substance of the glass forming the tube in which it is kept, either dry or in solution.

Lastly, and this is perhaps the most striking of the many results of the energy put forth by radium, it can produce very definite effects, both beneficial and destructive, upon the tissues of animals and plants. As this matter will have to be referred to more fully later on, the detailed description of its action on the human body will not be given here. It is sufficient for the present to call attention to the fact that the continued activity of radium is shown by physiological, as well as by chemical and physical effects.

In all these ways radium shows that it is constantly giving out a considerable quantity of energy; and yet we cannot find that, after working in these ways for a definite length of time, it has suffered any diminution or change. We appear, then, to have in radium an exception to the great law of the conservation of energy, which was explained above.

The other great natural law referred to is that of the persistence of matter; and to this also radium appears to offer an exception.

Before the days of exact science, it was not thought impossible that things which disappeared from the sight might be actually annihilated. Also, conversely, that things might come into existence out of nothing, or, putting it in another way, that a definite weight of material might, by special treatment, be turned into a greater weight of other material. We are now convinced that no such event occurs in the course of nature. If a pound of salt is dissolved in ten pounds of water, and so disappears from sight, we confidently expect to have eleven pounds weight of solution. If an ounce of pure charcoal is completely burned in an enclosed room, we know that it exists as really and substantially, though now invisible, as it did before. Part of the invisible oxygen in the room has been converted, by addition of the carbon, into another invisible but heavier gas, carbonic acid; so that the air of the room weighs exactly an ounce more than it did before the burning. On the other hand, if a cloud appears and grows in the blue sky, we know that its material has not come into existence out of nothing; but the moisture was there before, and has merely changed its state in such a way as to become visible. We have, in fact, recognised the great natural law or truth of the persistence of matter: no substance, to our knowledge, ever comes into existence out of nothing, and no substance is ever annihilated or changed into nothing.

Now it was soon found that if a current of air were allowed to pass over radium, it carried away something that could display the properties of radium itself, making willemite and other minerals luminescent, and ionizing the air in its neighbourhood, so as to make it a conductor of electricity. This material, which is carried away from radium by a current of air, can diffuse through gases and porous material, can be condensed by cold and re-evaporated by warmth, and can, moreover, be deposited upon the surface of any material object exposed to the air which contains it, and can afterwards be removed from that surface by friction or solution, and still display the properties and powers characteristic of radium. It is clear, then, that a portion of the radium itself has been carried away by the current of air. Some material, therefore, is being all the time exhaled or emanated from the radium. The material so given was called, and continues to be called, the emanation, and, apparently, its production never ceases or varies in quantity.

The astonishing thing is that, after years of this emanating work—after giving out for years a constant stream of substantial material into the air about it, the radium itself does not grow the smallest perceptible fraction of a grain lighter than it was before. That is to say, after a part of it has kept passing away for years, there appears to be none less left behind. It resembles the widow's cruse, from which successive supplies of oil could be drawn, without any diminution in the stock. This behaviour of radium, when observed, was the most obvious approach to a miracle that it has fallen to the lot of modern science to examine. Here was, apparently, an exception to the second great law of nature, the persistence of matter. Here was matter, it seemed, coming into existence out of nothing, not simply changing from an invisible into a visible form. To the ordinary mind this is even more surprising than an apparent exception to the conservation of energy. For the estimation of energy is a difficult matter, only made possible by the results of modern scientific work. The Protean character of energy makes it very elusive. It may appear in the form of momentum, pressure, expansion, heat, electricity, magnetism, light, sound, and various forms of chemical and vital activity. From one of these forms it may be exchanged into another, through yet others as intervening stages. And in any of these forms the amount of energy involved in any phenomenon was, until lately, very difficult to estimate; and it was impossible, before the days of Joule, to compare with exactness the amount of energy manifested in two different forms—to state the amount of heat, for instance, that contains a quantity of work equivalent to the effort of lifting a 20 lb. weight 12 yards high. But man has been accustomed, since the earliest days of civilisation, to estimate and compare quantities of matter, by the simple process of weighing. We therefore easily assimilate the doctrine that, since the practice of exact observation has prevailed, matter has not been known to come into or go out of existence: and so we can readily appreciate the exceptional or miraculous nature of a phenomenon which involves the continual production of material emanations, without any observable diminution of the matter in the source from which they are produced.

Here, then, we have the main facts which have given to radium and its phenomena a unique interest, which have made it a puzzling exception to the widest and most firmly established scientific generalisations, and have given it a hold on the imagination of the public, both indirectly on account of the scientific difficulties connected with it, and directly because of the wonderful possibilities in the practical and industrial world associated with its future development.

The puzzle has not long remained a puzzle. It has been solved in a manner which may prove finally satisfactory; and, at any rate, the explanation offered is such as to avoid a conflict with any of the well-established laws of nature, as we are accustomed to call them, that is, the systematically verified doctrines of science. But before we consider this explanation, it will be well to examine more fully the properties of radium, and the way in which they came to be discovered.

II.

The Discovery of Radium.

In recent years our knowledge of various kinds of radiation has been remarkably extended. We have long been familiar with vibrations of water, air, and ether, under the names of waves, sound, heat, and light. Hertz also investigated other ethereal vibrations, of an electric nature, which have recently been turned to great practical use by Marconi and others in the development of wireless telegraphy. In 1895 Professor Rontgen astonished the world by discovering another very surprising kind of radiation, which is emitted by an active Crookes' tube. A Crookes' tube is a tube of glass sealed at both ends, with the poles of an electric circuit welded into and passing through the glass to the inside of the tube, as at *c*, Fig. 11. The air or gas in the tube has been very highly exhausted, that is, very nearly all has been pumped out before the tube is sealed. When an electric current is passed through the wires, the small remaining quantity of gas in the tube is broken up into very fine particles, which fly from the cathode, or negative pole, with a very high velocity, and strike the opposite end of the tube so forcibly as to produce light, or make the glass fluorescent. The small particles of decomposed gas flying from the cathode, whose sudden stoppage produces the illumination of a Crookes' tube, are called the cathode rays, and it is to be noted that we are now dealing with rays of a very different kind from the radiations mentioned above. In the case of waves, or sound, or light, the radiation is the movement outwards of vibrations, while the vibrating matter remains in much the same place. The cathode rays are not moving vibrations of stationary matter, but moving matter—

a rushing stream of minute particles of disintegrated gas. When these small flying particles of gas are suddenly stopped, either by the walls of the tube or by an obstruction purposely placed inside the tube, the violence of the sudden stoppage not only produces light, that is, a continuous succession of regular or timed vibrations of the ether, able to excite in our eyes the sensation of light; but it produces also an irregular succession of untimed vibrations of ether, bearing to light much the same relation as a sustained noise bears to a musical note. These latter vibrations do not excite the nerves of vision. But just as an invisible electric current can excite in a filament vibrations which produce visible light, so these irregular vibrations from the Crookes' tube, though themselves invisible, can excite in certain substances, such as platino-cyanide of barium, vibrations of a kind which excite light-waves in ether, and so render the platino-cyanide visible whenever it is acted upon by the irregular waves from the Crookes' tube. These irregular waves were the famous discovery of Professor Rontgen, and accordingly they are called Rontgen rays, as well as X-rays. Röntgen found that these rays, invisible themselves, could not only make certain substances visible by causing them to give out light rays, but that they could do this after having themselves passed through substances which are opaque to light, such as wood, cloth, and flesh, as well as thin plates of metal; while denser materials obstruct them, generally in proportion to their density. The Röntgen rays can not only make certain substances luminescent, but can act as light does on a photographic plate: and here, again, the local intensity of the action depends upon the density of any objects through which the rays may have to pass; so that, if we employ a photographic plate, we get an X-ray shadow picture or skiagram of an interposed hand, limb, or other part of the body.

Another observation connected with the behaviour of substances in relation to light brings us still nearer to the discovery of radium, to which it led up. As barium platino-cyanide can receive invisible X-rays, and give out visible light rays into which it has transformed them, some substances have the property of being able to receive light rays, and give them out again not immediately, but after a longer or shorter interval, still as light rays, but slightly altered, as ordinary substances can store up heat. Calcium sulphide, specially prepared, is used to make a luminous paint which absorbs light during the day, and gradually gives it out again through many hours at night. So we have match-boxes with a cover which shines in the dark, provided it has been previously exposed to light. Some substances which have this power of phosphorescence after exposure to light, become much more rapidly exhausted. Professor Becquerel devised an ingenious piece of apparatus for measuring the duration of this phosphorescence after exposure to light, even when it lasted no longer than a small fraction of a second. He found that compounds of uranium exhibit a very bright luminosity for the thirtieth part of a second after they have been illuminated. We have seen that in the Crookes' tube the production of visible rays of fluorescent light was accompanied by the production of invisible X-rays which could pass through materials opaque to light, and then act on a photographic plate. It was suggested that all fluorescence might in the same way be accompanied by X-rays. Becquerel in 1896 tested this point with fluorescent compounds of uranium. He covered up a photographic plate with opaque material, so that it was safe from the action of light. Then he exposed a salt of uranium to sunlight, so as to make it fluorescent, and afterwards laid it on the photographic plate, with the opaque material still between—wood, paper, or very thin copper or aluminium. The plate, when developed, showed that it had been acted upon by the uranium; not because of its phosphorescence, for the light was stopped by the opaque layer, but under the influence of some rays that were capable of passing through the opaque layer, and were in that respect, as well as in their action on the plate, similar to X-rays. One day, however, one of those accidents happened which have so often proved useful to great scientific discoverers who have had the acuteness to observe them, and the knowledge to turn them to good account. The sun on one occasion became overclouded just as Becquerel was going to expose the uranium to sunlight, so that he could not proceed with the experiment. He decided to wait for another opportunity, and putting away the plate and the uranium, shut them up in a drawer. Another day, opening the drawer, and finding the plate with the uranium lying upon it, and the usual opaque screen between them, he developed the plate in the ordinary way, and got an impression of the uranium. Remembering, then, that the uranium had not been bathed in sunlight before being laid away on the plate, he reflected that this impression was more than he had a right to expect, on the theory on which he had previously been working. Further experiment, carefully directed to the point, showed that the theory hitherto held must be wrong. For, repeating the accidental arrangement, that is to say, using uranium without first bathing it in sunlight, Becquerel found that he again got an impression from it on the plate. More surprising was the fact that if he took a uranium salt which had been chemically made out of its constituents and freshly crystallised without ever in its existence having been exposed to light, this uranium salt proved to be equally capable of acting on a photographic plate through intervening opaque screens. It was clear, then, that light had nothing whatever to do with this remarkable action. The uranium itself, in virtue of being uranium, and without borrowing from any other source, as calcium sulphide does from daylight, possessed the power of emitting some kind of rays, which, like the X-rays from a Crookes' tube, can pass through opaque substances, and then act on a photographic plate. Here, then, was the first discovery of radio-activity, a wonderful property, the possession of which, in a pre-eminent degree, gives to radium its special interest and importance.

This property of radio-activity, the power of spontaneously, without known chemical change, and without known external help or stimulus, sending out invisible energy in such forms as to be capable of passing through substances, and producing chemical or other action at a distance, was investigated by others than its discoverer; among them by Schmidt and Madame Curie, who both discovered independently, in 1898, that of all the other elements then known, there was one, and one only, which possessed this remarkable property, namely, thorium. Mme Curie, the wife of Professor Curie of Paris, a Polish lady by birth, made a special study of radio-activity, and carefully examined a large number of minerals to find which of them manifested this property in the highest degree. It was very naturally looked for in pitch-blende, the one from which uranium itself is obtained. Pitch-blende is obtained largely from Austrian mines, and it varies in the degree of its radio-activity. That obtained from the Johanngeorgenstadt mines, which was found by Mme Curie to be the richest in radio-activity, contains 70 per cent. of uranium. Pitch-blende consists largely of oxides of uranium and iron, but it contains also as impurities lead and most other metals, besides a large number of other minerals. If the radio-activity of pitch-blende from Johanngeorgenstadt be indicated by the number 8 3, that from another Austrian mine, Joachimsthal, was found by Mme Curie to be 7. Pitch-blende is also found in other places, among them Cornwall; but the radio-activity of the Cornish pitch-blende was found to be no higher than 1·6. Among the other minerals in which radio-activity was observed is cleveite. Cleveite was already famous as the source from which helium was first obtained, and helium, as we shall see later, has a very close connection with the most radio-active of all substances, radium.

It is interesting here to consider the means by which the radio-activity of a substance was estimated, for the method employed is an illustration of the utmost refinement yet reached by science in the measurement of small quantities. By this means it is possible to detect the presence of substances in quantities thousands of times as small as could be weighed by the most delicate balances, or measured by the aid of the most powerful microscopes, quantities far too small to be identified even by means of spectrum analysis. It has been mentioned that dry air, which is not a conductor of electricity, can by appropriate means be broken up in such a way as to make it a conductor. This is called ionizing the air. The vibrations of ether, with which we have been familiar under the name of X-rays, can not only act on a photographic plate, but can also ionize the air through which they pass. It was soon found that the action of uranium, thorium, and radio-active minerals resembled that of the X-rays in the latter respect as well as in the former. They also can convert the air in their neighbourhood into a feeble conductor of electricity, with more or less completeness, according to the amount of their radio-activity; and electric measurements can now be made with such extreme refinement, that it is possible by this means to detect infinitesimally small changes in the conductivity of the air, and therefore in the radio-activity of the substances which are making it a conductor. The delicacy of observation made possible by this means is far greater than can be attained to in observing photographic effects. Moreover, the observations can be much more rapidly, as well as more conveniently, completed. In the case of skiagraphic work, many hours of exposure would be required with feeble radio-actors, in order to get a visible impression, whereas in many cases these phenomena undergo such rapid change, that observations, to be of use, must be made in a few seconds. And, to say nothing of the time and care required in developing, the comparison of impressions in regard to their strength cannot be made with

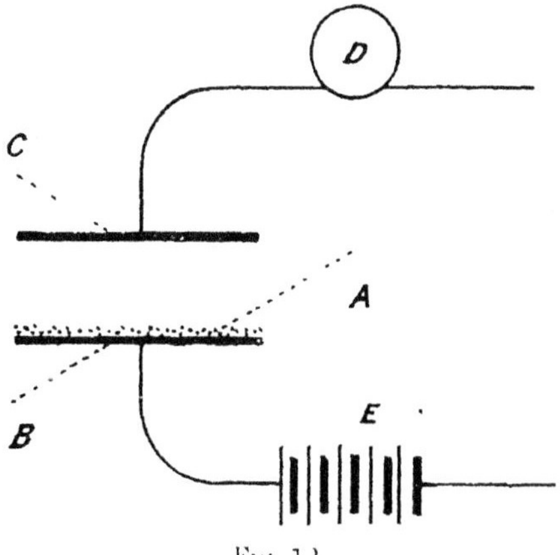

Fig 12.

any approach to the extreme delicacy that is easily realised in comparing the strength of electric currents. Accordingly ionization, rather than skiagraphic action, was the manifestation of radio-activity chosen by Mme Curie for observation in detecting and comparing the radio-activity of various substances. The essential part of the apparatus which she employed is shown in Fig. 12. The material to be examined is powdered and spread out in a thin layer A over a flat metal plate B, of about 6 square inches surface. About 2 inches above it is another similar plate C. To the plates are attached wires forming parts of an electric circuit, and passing through a battery E to generate current, and a buadrant electrometer D to measure its strength. The rest of the connections are not shown in the figure. Air being a non-conductor, the circuit is broken between the two plates, and no current can pass. But if the substance in the layer A is radio-active, it ionizes the air immediately over it between the two plates, making it able to conduct electricity. The ions produced in this air can carry charges of electricity, and so it is now possible for a current to pass. The amount of current passing increases with the rise of the voltage up to a certain maximum, called the saturation current. A stronger radio-active substance on the plate B would produce more ions in the air, and so allow a stronger saturation current to pass. Thus the amount of current passing is a measure of the radio-activity of the material A. This current is itself measured by the charge it gives to one pair of quadrants in the electrometer D, which would move the electrometer needle, were it not balanced by a similar and opposite charge. The latter charge Mme Curie obtained by putting a quartz lamina under tension, with whatever weight was just sufficient to produce the neutralising charge. This weight became, therefore, a measure of the neutralised charge, and so of the current producing it by passing between the plates, that is to say, of the radio-activity of the material which makes the passage of the current possible. Professor E. Rutherford, of the M'Gill University, Montreal, who has done a great deal of important work in elucidating the properties of radium and similar substances, measured the charge in the electrometer more directly by observing the time taken by the electrometer needle in moving over a certain number of divisions in the scale: but as the indications of these delicate instruments are liable to errors arising from many sources, such as the state of electrification of objects in the neighbourhood, it is necessary to guard against these by taking many elaborate precautions, which, however, need not be described here. It is sufficient to recognise that apparatus of this kind affords a means of making excessively delicate comparisons between the degrees of radio-activity of different substances, as indicated by the extent to which they produce an ionizing effect upon the air in their neighbourhood.

Armed with such apparatus, Mme Curie, as we have said, examined a large number of elements and minerals. It was found that uranium displayed its radio-activity whether it was tested as the separated metal, or in compound forms combined with other substances, as in the oxide or sulphide, or salts of sodium, potassium, or ammonium. The radio-activity of the compound substance was usually in proportion to the richness of the compound in uranium. The uranium was, as we have seen, extracted from pitch-blende, the best Austrian pitch-blende containing it to the extent of 70 per cent. It was wanted largely for colouring glass. The remaining 30 per cent., now known as uranium residues, were rejected as worthless. The most amazing result of Mme Curie's examination of the radio-activity of different substances was the discovery, that although it was the behaviour of uranium that had revealed to Becquerel that there was in the world such a thing as radio-activity, still uranium was surpassed four times over in the intensity of this property by the pitch-blende from which it came. The activity of the uranium oxides (black and green) being marked as 2·6 and 1·8, that of pitch-blende was found to be 8·3. The pitch-blende was stronger in radio-activity than the uranium. And, inasmuch as the residues left after removing uranium formed but a small part of the whole mineral which they made so much stronger than uranium, these residues themselves must be immensely more active than uranium. But they consisted largely of iron, lead, and a number of other substances which were known to have no radio-activity at all. It followed that they must be mixed with a small quantity of some yet undiscovered substance, which was the cause of the observed activity; and this material, being so small in quantity, must be of an activity correspondingly intense, in order to leaven the whole lump so effectually as it did.

The search for this small quantity of hitherto undiscovered material is one of the most remarkable pieces of scientific work in verification of a previous train of reasoning. In the actual work, Mme Curie was assisted by her husband Professor Curie and by M. Bémont. The chemical analysis of the uranium residues is described by Mme Curie, but a repetition of it would be meaningless for our present purpose. A very brief summary of it occupies a large page and a half of print, and it must suffice us to say that it is a very long and tedious process, requiring many months for its completion. After each step which separated one group of substances from another, the investigators compared, by the electrometer test described above, the two groups, for the purpose of determining which of them possessed the greater amount of radio-activity. This one would be supposed to contain the substance, or the greater proportion of the substance of which they were in search, and it would be further subdivided into smaller groups to still further narrow the limits within which the object of their search must be looked for.

At one stage in the analysis, the precipitate obtained by adding sulphuretted hydrogen to an acid solution, gave, among other things, a form of bismuth which was found to be very radio-active. But bismuth itself, obtained in other ways, is not radio-active. The inference was, that, mixed or combined with the bismuth, and separated by the same chemical reactions which had separated the bismuth, was another substance, strongly radio-active. This proved to be the case, and the radio-active material was further concentrated by processes of sublimation and precipitation, proving to be more volatile than bismuth as a sulphide, and less soluble as a nitrate or a sulphide. Still, in its chemical behaviour it so strongly resembled bismuth, that Mme Curie found it impossible to get it quite purified from that element. In honour of her native country, the new substance was called polonium. Marckwald has since devised another method by which it is got purer, though still not quite pure. The extreme delicacy of these researches, dealing as they do with excessively small quantities, is well illustrated by the fact that by his improved method Marckwald obtained out of two tons of pitch-blende one sixteenth part of a grain of polonium—one part out of five hundred millions.

The bismuth portion of pitch-blende was not the only portion which Mme Curie found to be radio-active. Just as polonium imitated very closely the chemical reactions of bismuth, so it was with barium and some other substances. The barium obtained from the pitch-blende by ordinary methods of analysis appeared to be very radio-active. But since ordinary barium is not radio-active, here again it was inferred that some radio-active substance, behaving chemically in very much the same way as barium, had been separated with it, and means were sought of getting it by itself. It was found that the chlorides of barium and the radio-active substance, if dissolved in boiling water and allowed to cool, crystallise out of solution in such a way that the less soluble portion proves to be more active than the remainder. If this more active portion is treated in the same way, it is itself divided into two portions, of which the less soluble is still more active than before. Thus, by taking advantage of the fact that the radio-active material is less soluble than the barium chloride, a gradually increasing concentration of the former can be obtained by continually repeating this process with the most concentrated portion yet obtained. Such a method of getting a gradually more complete separation of two substances which are almost exactly alike under ordinary chemical tests, is called a process of fractionation, in this instance fractional precipitation. By pursuing to the end, with the greatest skill and perseverance, this method of increasing purification, Mme Curie succeeded at last in separating from the barium of pitch-blende a very minute quantity of a substance which, closely as it resembled barium in most ways, differed from it completely in the possession of an enormously high ratio-activity—an activity a million times as great as that of uranium, from which the first lesson had been learned of this particular class of phenomena. Very appropriately, therefore, she christened the new substance radium. She had it sufficiently purified to give an indication of its individual spectrum in 1898. Radium also is found in very small quantities, though not quite so small as those of polonium. Mme Curie calculated that from two tons of uranium residues of pitch-blende, she obtained about three-quarters of a grain of fairly pure radium chloride —one part in forty-two millions. Giesel has the radium in the form of the bromide salt, instead of the chloride, when he comes to the fractionation point in the extraction, and with this method he considers that he gets better results. He gets out of two tons of pitch-blende residues nearly eight grains of fairly pure radium bromide—one part in four millions.

A fifth radio-active substance was in 1899 announced by Debierne. This was also extracted from pitch-blende, but separated in another part of the chemical analysis. It appears to be the same substance as was observed by Giesel, and described by him as the emanation substance. Debierne called it actinium, which is derived from a Greek word, and means the radiating element. Radium, from a Latin word, has the same meaning.

III.

The Radio-Active Elements.

We are thus in possession of five elements which possess the property of radio-activity; but as this is a composite property, including various manifestations different in detail, it will be well, before considering radium more particularly, to compare with it the other four as to the nature of their radio-activity. It was mentioned that radium constantly gives out heat and ionizes the surrounding air, which effects are chiefly due to the action of what are called α rays. These rays, carrying a positive charge of electricity, can be made by a magnet to deviate slightly from rectilinear motion, and are easily stopped by a thin sheet of metal, or even by a piece of paper, or an inch deep layer of air, which they exhaust themselves in ionizing. These α rays are possessed by all the other four radio-active elements. Radium also sends out rays which are chiefly manifested by photographic action. These, the β rays, are very much like the cathode rays of a Crookes' tube. They carry a negative charge of electricity, are very deviable in a magnetic field, move with a much higher velocity than the α rays, and readily pass through cloth, paper, and wood, and even thin sheets of metal, especially the light metal aluminium; but are practically all stopped by a dense metal equal in thickness to a penny piece. The β rays are given out by all the other radio-active elements except polonium. The α and β rays will all be stopped by a sheet of iron of the thickness of half an inch. But a third kind of rays from radium, the γ rays, will get through. They have such an extreme power of penetration, that their action can be detected even after they have passed through an inch of lead or twelve inches of solid iron. In their passage they cannot be deflected by a magnet. Great as is their power of penetration, their effect is feeble compared with that of the preceding rays. They greatly resemble the X-rays of Rontgen, and manifest themselves by photographic action, by ionization of air, and by exciting luminous effects on a screen of platino-cyanide of barium or the mineral willemite. These γ rays are emitted by uranium and thorium as well as by radium, but not by polonium;

and the case of actinium is still doubtful. It was seen that radium, besides emitting these three kinds of rays, gives off into the air some material, called emanation, which behaves in many ways like a gas, and which manifests its presence both by exciting luminescence in suitable screens, and also by depositing upon any surfaces with which it comes into contact some material which enables them, in their turn, to display the possession of radio-activity, as in the ionization of air. Such activity of a secondary kind, derived from the presence or passage of emanations, is called excited activity. This radio-active emanation and excited activity, observed from radium, are observed also from thorium and actinium, but not from uranium or polonium.

Such are the five known radio-active elements, uranium, thorium, polonium, radium, and actinium. Their distinctive identities are not, in all cases, quite finally cleared up. It is, though probable, not absolutely certain, that a substance discovered by Marckwald, and named by him radio-tellurium, is identical with Mme Curie's polonium. It is most likely, also, that Giesel's "emanation substance from pitch-blende" is the same as Debierne's actinium. It must be remembered that the quantities of these substances under observation are excessively small, and that they are in very few hands, so that it is not possible for anyone interested in the matter to step in and test the few observations made, with a view to confirming or correcting them.

The pure metal radium is never seen. It has always been obtained in combination, in one of its salts—the bromide, chloride, acetate, sulphate, or carbonate. The metal itself would probably prove to be very unstable, and oxidise rapidly in the air, like sodium and potassium. Chemically, it behaves in very much the same way as barium, and must be classed with magnesium, calcium, and strontium, as one of the alkaline earths. Its chloride and bromide form crystals which are self-luminous in the dark. They are a clear white at first, but become brownish with the lapse of time. The atomic weight of radium is 225, as estimated by Mme Curie. This makes it the heaviest element atomically, with the exception of two, uranium being 238 and thorium 232, and both of these also being radio-active elements. The quantities in which polonium and actinium have been observed are too small to admit of experiments for determining their atomic weights, so that, up to the limits of our knowledge, the radio-active substances are those which have the highest atomic weight.

IV.

The Miracle of Radium.

When Becquerel discovered that the power of uranium to act upon a plate through a piece of wood was not a restoration of energy borrowed from the sun, but an inherent property of the uranium, he had no connected theory to offer in explanation of its action. When the Curies observed the same properties in thorium, polonium, and radium, and found that they all possessed also the power of ionizing the air, still there was no reasonable account of these strange happenings to be given. Here were substances perpetually working without help or stimulus, yet never exhausted, a thing out of all order and reason, so far as our previous knowledge went of nature's laws. These manifestations of energy, however, were on so small a scale, and so far from obvious, requiring scientific arrangements to render them apparent, that they did not clamour for attention. The question how they were produced, admitting, as it did, of no immediate solution, admitted, nevertheless, of being left for a while in suspense, in the faith that an explanation in accordance with nature's ascertained laws would probably be found in reasonable time. But when the Curies found that radium was constantly giving out heat, without ever getting cold or being burnt away, the case was very different. The phenomenon at once became a striking puzzle, pressing for solution. If, at the end of a journey to Aberdeen, we found the foot-warmer in our carriage as hot as on starting from London, without apparent loss or change of material, we should want very much to find out how it was. That was the attitude of the scientific world towards the amazing fact that radium kept giving out heat, without assistance, and without apparent loss or change. It was a thing without parallel in the whole experience of research since modern science began. It seemed to require a re-examination of one of the widest and safest scientific generalisations—that of the conservation of energy. At the British Association meeting of 1903, distinguished physicists expressed the view that if such a strange thing really happened, it was not merely a mystery, but a miracle. A mystery is a puzzle unsolved, though not necessarily incapable of solution. A miracle is in flat contradiction to known rules, and this was the case of this constant expenditure of heat without loss. A suggested explanation that was favoured for a time, was that radium had the power of indefinitely transforming energy. We may see a piece of iron beaten until it becomes white-hot. The energy of the falling hammer is transformed, in the substance of the piece of iron, into the energy of heat—merely another kind of movement. So, it was thought, radium may have the special power of transforming some unknown form of energy, which it finds always and everywhere present, into such well-known forms as heat, light, electric and photographic action. The puzzle was to say what this unknown form of energy could be.

Moreover, it is impossible to suggest in detail the mechanism by which energy could be converted into those forms in which it is manifested by radium. Accordingly there were many, and among them the Curies themselves, who were not convinced that the attempt to explain the phenomena of the radio-active bodies was on the right lines. The phenomena were studied with great minuteness and originality by Professor E. Rutherford and Mr F. Soddy at the M'Gill University, Montreal. They made most of their early observations upon thorium, the action of which is in many ways like that of radium, though very much less energetic. Their studies led them to form, apparently in 1902, and they published in the *Philosophical Magazine* in 1903, a connected and systematic account of an explanation of the phenomena of radio-activity, which they devised for the manifestations of thorium, and afterwards applied to radium and the other radio-active bodies. It depended on the view that, whereas probably all atoms consist of a large number of much smaller parts, the corpuscles; the radio-active bodies constantly have some of their atoms disintegrating, that is, breaking up into their constituent corpuscles or groups of corpuscles. This was revolutionary indeed; and to many minds such an explanation was even more unwelcome than the other. For just a hundred years it has been, if not a main article, nevertheless one of the best established articles of scientific faith, that the atom is the smallest separable part of matter, and also that it is indestructible. Its name atom, derived from the Greek, signifies indivisible, and none but a few in recent years have dreamt of the possibility of its ever being divided or having its parts re-arranged. Scientific men of the very highest standing, regarding the atoms as the smallest particles that research would ever discover to represent the material out of which things are built up, did not hesitate to call them the foundation-stones of the universe; and now, in explanation of the phenomena of thorium and radium, a theory is put forward which teaches that there are smaller things than atoms, and that atoms themselves, imperishable atoms, are constantly breaking up. To the more conservative minds this explanation was altogether repugnant. Such a revolution in the details of Dalton's atomic theory, as they and their teachers had believed in it for a hundred years, seemed unthinkable. Professor Armstrong, in a communication to the Royal Society, while not dogmatically denying the new doctrine (few scientific men now venture to deny anything dogmatically), nevertheless ridiculed the idea of the atoms breaking up in the manner suggested by this theory —"which assumes," he says, "that nature has endowed radium alone of all the elements with incurable suicidal monomania." This view has now, however, commended itself to the great majority of the scientific world as the true explanation, and we shall call it the true explanation in the rest of this book, in which it will be set forth more fully.

We can hardly admire too much the skill in discovery of Becquerel, the Curies, Debierne, and Giesel, who have shown the instinct of a hunter in scenting new truth, and a hunter's keen observation and patience in tracking down and presenting to the world the most puzzling discovery that it has received since science began. We cannot admire less the painstaking and persevering researches into all the detailed facts of this puzzle, which enabled Rutherford and Soddy to develop the ingenious suggestion which alone has been found equal to giving a satisfactory explanation of all those facts. Others have since been very busy protesting, in the name of science, that there is nothing in it all to be much astonished at. We are told that there are no facts in these discoveries which contradict the established laws of nature as previously known to science. That if they revolutionise some scientific views, they do not revolutionise science. That scientific men were quite prepared for all that has been discovered, and that every new fact fits into its place in the old theories. That, in short, science has a layette ready prepared to receive every new discovery the moment it is born. The inference is, that the "wonder-mongers," who have been holding forth on the marvels of radium, are making a great disturbance over the discovery of what will prove to be merely a mare's nest.

Let not the amateur student of science be discouraged; and let him not be ashamed of the surprise with which he first heard of the properties of radium. Plenty of distinguished physicists were willing to declare, no longer ago than last year, that the facts about radium, if verified, were more than they could harmonise with the laws of nature, as they understood them. The President of the Section of Mathematics and Physics at the 1903 meeting of the British Association was Professor Boys, who said, in his opening address:—"The discovery by Professor and Madame Curie of what seems to be the everlasting production of heat in easily measurable quantity by a minute amount of a radium compound is so amazing that, even now that many of us have had the opportunity of seeing with our own eyes the heated thermometer, we are hardly able to believe what we see. This, which can barely be distinguished from the discovery of perpetual motion, which it is an axiom of science to call impossible, has left every chemist and physicist in a state of bewilderment. This mystery is being attacked, and theories are being invented to account for the marvellous results of observation; but the theories themselves would, a few years ago, have seemed more wonderful and incredible than the facts, as we believe them to be, do to-day. With all this mystery before us, which I must confess myself wholly unable to follow——." And Sir Robert Ball, in a recent article on radium, speaks of—"the mystery of this substance, to say nothing of the miracle;" and of—"the profound impression made by the discovery of radium. It marks an epoch in the history of our knowledge of nature." With such words from such men in his ears, no one need fear to be called stupid if he listens with opened-mouthed astonishment to the tale of the marvels of radium, or to be called a wonder-monger if he repeats them. Neither need he accept the myth that science has experienced no revolutionary change, because it has always a layette ready for every new discovery. The true view of the ordinary course of scientific progress is that given in Professor Boys' account above of its development in the case of radium,—the new and surprising facts are discovered first, and then the existing theories are modified or given up to suit them. If all was so easy to understand, how is it that there were distinguished chemists, who, in the words quoted above, found the theory offered in explanation of these facts—a theory now generally accepted as true—so far beyond belief as to be actually ridiculous? When the first of these facts was discovered by Becquerel in 1896, the gentlemen to whom, as representatives of science, nothing is surprising, had not a word to utter in exposition of that theory which they now accept. The child was born, the layette, as they now tell us, was all ready and well known to them, but they kept its existence and its whereabouts a secret from the nurse, till Rutherford and Soddy discovered, or rather made, it in 1902. The truth is that the layette was not ready for the new facts. The cloth and flannel were there, the scissors and needle and thread were ready; but the making up had all to be done. It was after the new facts were discovered that, to explain them, new theories were developed and old theories abandoned. This is the natural and ordinary course of scientific progress. Which of these unastonishable physicists had predicted, before Becquerel's discoveries in 1896, that within a few years a substance would be found which was capable of giving out heat continuously for years, without apparent loss or change? We may be very sure that if to any of them such a thing had been foretold, he would have scornfully declared it impossible. *Ex nihilo nihil fit.* It is the realisation of this impossibility, the apparent creation of energy and matter out of nothing, which has so profoundly astonished the world, and that chiefly because of the profound and very evident, if temporary, astonishment of science. The various details of the ways in which the miracle manifests itself, and the accepted explanation of the means by which its causes operate, are the subjects of the remaining pages.

V.

THE STRUCTURE OF MATTER.

It is possible for a very correct account to be a very poor explanation. To one who had never played a game of cards of any kind whatever, or even so much as seen a pack, you could not, by reading the rules of bridge, give any comprehension of the real nature of the game, or the interest aroused by it. So to the ordinary layman, who has no knowledge of physics or chemistry, beyond such acquaintance with their application in practical life as he gains from newspaper research, or from experimentation with pills and telephones, a technical account of a scientific subject conveys no illumination whatever. The unscientific reader finds almost every page of a technical book on such subjects bristling with algebraic formulæ, of which he understands none, and with scientific terms of which every other one at least is meaningless to him, while the treatment of the whole subject takes for granted in the reader a familiarity with modern scientific conceptions as a foundation on which the whole discussion rests. This foundation is not present in the ordinary reader's mental equipment, and therefore an explanation, to be adequate for him, must provide a sufficiency of foundation work before commencing on the main structure. The building so reared will necessarily be smaller and less elaborate; but, so far as it goes, it will be solid and substantial. The explanation will be a real one. A true understanding of the properties and actions of radium involves a knowledge of the way in which its constitution and behaviour in minute detail resemble and differ from those of ordinary substances. This requires a knowledge of the minute structure and action of matter in general, and that not merely in the sense of reading a written account of it, but to the extent of getting, by illustrations and examples from similar phenomena on a larger scale in common experience, such a familiarity with what is going on all round us in things of far less than microscopic size, that the imagination can take, as it were, an inside view, and so obtain a more sympathetic, and, therefore, more complete understanding of the doings of the almost infinitely small existences with which this study is concerned If the mind does not get into such a familiar and friendly attitude towards those minute bodies, the tales we have to tell about their behaviour will prove, to a great extent, unintelligible, and, therefore, incredible to those who cannot go through the physical and mathematical trains of reasoning which compel belief. Such a familiarity with the contemplation of the excessively minute is best obtained gradually, as, indeed, it is gradually that the knowledge of it has been won; and we shall, therefore, begin by looking cursorily into the structure of matter on a larger scale.

A study of the structure of matter brings out prominently at every stage, from the mightiest masses to the tiniest particles, two great pervading principles—subdivision into portions or groups, and relative movement of these portions. We will begin with matter on the largest scale on which we know it—the whole universe. A glance at the sky on a clear night suggests at first that the suns, stars, worlds, and other bodies which compose the universe, are distributed in a fairly uniform manner through all that portion of space which we are able to examine. Careful observation shows, stretched across the middle of the sky, an irregularly shaped band of something bright, which looks very much like mist or thin cloud; but it can be seen every fine night, by any one with good eyes, to be of exactly the same shape, and in exactly the same position among the stars. It cannot, therefore, be mist or cloud, or any terrestrial matter at all. It is well known as the Milky Way; and though it is clearly visible to the naked eye, examination with a telescope shows that what looks like star-mist to the eye, really contains millions of separate stars, so small, when seen at this great distance, as to be indistinguishable, and so closely packed together as to produce the appearance of cloud. The telescope, however, shows patches here and there which still resemble mist, though, in many cases, these also are shown by still more powerful telescopes to be really separate stars. The fact appears to be that our sun is merely one of many millions of stars, more or less like himself, which are grouped together in one great swarm or cluster. The shape of the cluster appears to be round, but very much flattened, something like that of a tea-cake, with a deep cut between the two flat sides. Now anyone standing in the middle of a long narrow wood of tall fir trees, if he looked towards the sides of the wood, could see large patches of field and sky between the widely separated trunks of the trees; but if he looked towards the ends of the wood, he would perceive the spaces between the near trunks partly occupied by the trunks of trees farther off, the spaces between these latter being occupied by others still more distant, so that the whole line of sight in those directions would be occupied by tree trunks. Similarly anyone looking from a central point within our flat stellar system, would see, in certain directions towards the sides, merely stars or groups of stars, with spaces of dark sky between. But looking towards the edges of our cake-shaped swarm of stars, through the uncounted millions of miles of intervening star-filled space, he would see between the nearest stars the light of those more remote, between them the rays from others more distant still, and so on. The result is that, looking from inside towards any part of the outer edge of our great star-system, we find the whole field of sight in those directions covered with an apparently continuous layer of star-dust, visible all the way round the edge of the cake—the ring which we speak of as the Milky Way. The Milky Way, then, is the revelation to our naked eyes of the fact that the universe of matter is not evenly distributed through space, but is, in our neighbourhood, a bounded and limited and definitely shaped group of suns and stars—our stellar system. Astronomers have found, by the aid of powerful telescopes, that the universe contains other stellar or nebular systems also—some of them probably consisting, like ours, of millions of stars and suns, many of them, perhaps, with their own planets and satellites, comets and meteors, though the systems look, at their immense distances away, like faint splashes of hazy light,—others, perhaps, containing no stars at all, but only incandescent material something like gas. The distance of the other stellar systems from our own is so enormous, that the difficulty of discovering them is very great, and it is probable that the number of those yet undiscovered is many thousand times the number now known; that the universe is, in fact, divided into stellar and nebular systems in numbers comparable to those of the stars and suns within our own stellar system.

We find, then, that matter in what we will call the

first stage of existence, the universe, is divided into huge groups of substance, the stellar or nebular systems, separated from each other by vast distances of space, and probably moving rapidly through these spaces, though the motion is made apparently so slow, owing to the immeasurable magnitude of their distance from us, that no human means of observation can detect any movement at all.

Coming to the second scale of existence, the stellar and

Fig. 13.

nebular systems, we need not dwell on the division into stars or solar systems, as seen in Fig. 13: we cannot look at the sky without seeing the thousands upon thousands into which our own stellar system is divided. But on this scale we are able to verify what was a supposition in reference to the first scale, the state of relative movement between the component parts. The fact that the planets, or satellites of our sun, are continually changing their position among the stars, has been well known ever since the dawn of civilisation. But it is not the movements of the planets that we are now considering.

They are merely component parts of our solar system, our star; and at such distances as separate us from the majority of the stars, small bodies like the planets, not self-luminous, are completely invisible and out of the reckoning. The other stars, however, the true stars or suns, have been through all ages of human observation the most perfect embodiments of firm and unchanging stability. Planet is a Greek word, meaning wanderer; and on purpose to point the contrast with the planets, the other stars have been intentionally called the fixed stars, with reference to their freedom from motion relatively to one another. Religious teachers have never found anything better to point to than the stars as unchanging images of wisdom and truth. Lovers have sworn by them as models of faithful constancy. The science of all ages, seeking the most firmly fixed and permanent marks by which to scale the movements of sun and earth for the purpose of dividing time, has chosen for this purpose the fixed stars. Yet now we learn that the stars in our stellar system are constantly changing their position and arrangement. That the change has not until recently been detected by the most careful observation, is due to the enormous distances from which we see them. If we merely look out from a high hill over the sea and observe a fishing fleet on the distant horizon, the boats all appear at first to be stationary. It is only by long and careful watching that we can make out movements and changes of position. The fixed stars, as we call them, are so much farther off, in comparison with the extent of their movements, that the longest human life is not long enough to render their movements visible to the eye. There is reason, however, to think that historical time has been sufficient for an observable change in the shape of the constellation known as the Great Bear. The stars which compose it do not occupy exactly the same positions relatively to each other, which were plotted by Ptolemy two thousand years ago. The exact observations rendered possible by the refinements of modern instrumental work have led to the discovery of movements not previously suspected. Sir William Herschel found that our sun, with the earth and the other planets and attendant bodies, the whole solar system, is moving at the rate of five and a half miles a second towards a spot in the sky where we now see the constellation Hercules. The star which is charted as 61 Cygni moves through space at the rate of thirty miles a second. Another star, known as 1830 Groombridge, has a speed of two hundred miles a second. It thus appears that the millions of stars in our great stellar system, far from being the perfect model of fixed stability which we have so long been accustomed to think them, are much like a swarm of bees which have not yet settled into a cluster, and still appear as thousands of individuals flying freely about one another in the air. The reason why the corresponding movement of the stars in our star system is not perceptible, is simply that, as explained above, the distances of space and periods of time involved are so enormous in comparison with our limited base of observation and our brief span of existence. Passing for a moment to the consideration of nebulæ, whether in our stellar system or external to it,—the very shape of a nebula not seldom indicates a state of motion in its component parts. Spiral and radial formations—both of them, in our experience, implying motion—are found in the nebulæ. And in the case of some nebulæ, stars associated with and apparently helping to form some of their branches, have had definite movements relatively to the rest of the nebulæ observed and measured; so that in the second scale of existence, matter in the stellar and nebular systems, we find the two principles of subdivision into parts, and relative movement of the parts, clearly established.

In the third scale of existence—the stars, or solar systems—the same two principles manifest themselves unmistakably. Firstly, a solar system, such as our own (see Fig. 14), is divided into planetary systems, with cometary and meteoric systems, which are, perhaps, only variations of the former. Secondly, these systems move past one another and round the sun in the course of their orbital revolutions.

Similarly in the fourth or planetary scale of existence, a planetary system is divided into the planet itself and its attendant satellites or rings, as in the case of Saturn (see Fig. 15); and in this case also the component parts move past one another and round the central body. Planetary systems then illustrate once more the combination of the two principles of subdivision and relative motion.

In the fifth scale of existence we have the components of a planetary system,—single astronomical bodies. Of these, the best known to us is our own earth, which we readily perceive to consist of a large number of different bodies—the masses of the atmosphere above, the masses of water in oceans and lakes, the solid crust of many different kinds of rock and metal, and, in spite of some fashionable views which are not very well supported by evidence, the molten interior beneath the solid shell. So much for the subdivision of the earth into various masses of matter; and we have no reason to doubt that other astronomical bodies are constituted to some extent in a similar manner.

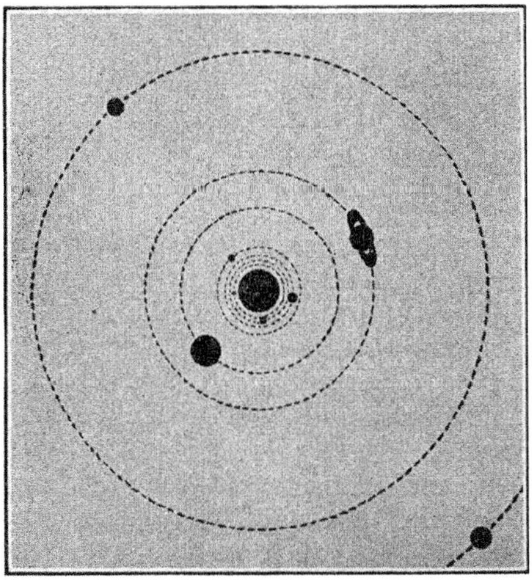

Fig. 14.

Then as to the relative movement of the component parts of an astronomical body. In the case of the earth we are familiar with the movements of the air in winds, of the water in tides, ocean currents, rivers, and rains; of the solid earth itself in earthquakes and in the elevation or depression of continents; while the extrusion of lava from volcanoes is fair evidence of movement of those molten masses beneath the crust, whose existence it has been one of the more recent affectations of science to doubt. The sun itself, to consider it for the time as a single astronomical body, shows vast relative movements of his component masses in the formation and changes of the sunspots, which appear to be mighty openings in his outer envelopes, displaying portions of the deeper masses. The moon, which is, with the exception of the earth, the easiest astronomical body for us to examine, appears at first to present an exception to this rule of movement of parts. We can see no evidence of any movement going on above, or on, or beneath the surface of the moon, though there have been some reports recently of the observation of superficial changes, as if the volcanoes were still at work. There is, however,

Fig. 15.

abundant evidence of movement in the past. Many parts of the earth's surface which are now the quietest and most firm, still display, in the structure and material of the mountain formations which have not yet been entirely worn away, the presence of numerous and mighty volcanoes in past ages. All over the moon, volcanic craters are the most obvious features of the surface (see Fig. 16), showing that our satellite cooled and hardened out of a condition in which the greater part of the crust was a sea of thickening lava, diversified only by rising heaps of scoriæ and volcanic dust, forming mighty craters 50 and 60 miles across. In the astronomical bodies, then, we have subdivision into masses, and relative movements of those masses well displayed.

We pass now to the examination of matter in the sixth scale, the masses of substance which go to make up an astronomical body. In our earth these are masses of gas, such as the oxygen and nitrogen of the air, masses of water in the seas, rivers, and lakes, masses of granite, sandstone, and a hundred other kinds of rock, masses of gold, copper,

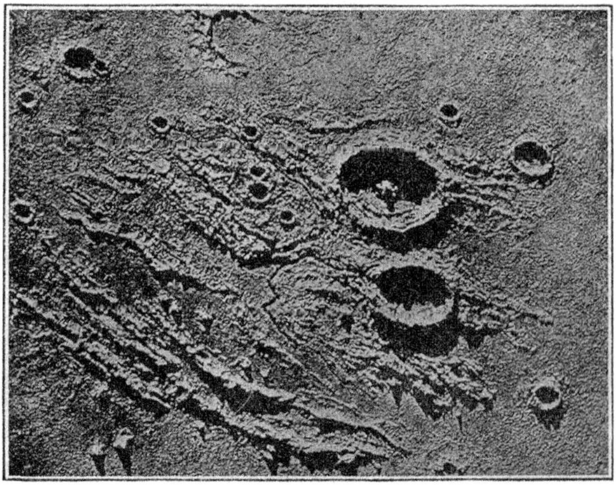

Fig. 16.

and numerous other metals; there are also organic masses, forming the bodies of animals and plants, and masses of organic products, such as limestone, coral, coal, peat, and petroleum. All these diversified objects are, in minute detail, constructed on one and the same principle. If we take salt or sugar and grind it up finely in a mortar, we can produce extremely small particles of dust, the smallest objects which can be seen in a microscope. But they are still particles of salt and sugar respectively, and a sufficient quantity of them would taste as such, and would give all the chemical reactions proper to those substances. By dissolving the material in water, we can divide it into still smaller portions. The smallest visible speck of salt or sugar, dissolved in twelve gallons of water, would be distributed through a million drops, and so divided into a million parts, of which one would be obtained in each drop of water. Yet here also the material would be the same salt or sugar with which we started, and, if concentrated sufficiently by evaporating the water, would again give the characteristic taste and chemical reaction. By chemical decomposition, however, and by heat, it is possible to break up the material so that we no longer have the same kind of substance with which we started. The salt or sugar has been so finely divided that we have got some particles smaller than the smallest existing particles of salt or sugar, namely, the elementary particles of which they are composed—sodium, chlorine, carbon, hydrogen, and oxygen, either separately, or recompounded with the particles of other elements. Let us, however, in our division, stop short of this point—the destruction of the material. Let us imagine it divided only till we have got the smallest portions in which it can exist in possession of its original properties. These smallest portions are all of one size and of one nature in each separate substance. They all contain exactly the same number of smaller particles of the component elements. Thus, each smallest possible portion of salt contains two particles, one of sodium and one of chlorine, and is exactly like every other smallest possible portion of salt. So with the smallest possible portions of sugar. If the sugar is all of one kind, each of its smallest possible portions is exactly like every other. They all contain the same number of particles of carbon, of hydrogen, and of oxygen. When the substance in question is an elementary one, such as gold or sulphur, each smallest possible portion contains, as a rule, two of the similar elementary particles. These smallest possible portions of substances which can be obtained without changing the nature of the substance, are called its molecules. The word is a Latin one, and simply means a very little bit. Everything that exists, solid, liquid, or gas, consists of molecules. Everything that we can feel or see, taste or smell, however smooth and continuous its substance may appear to us, would be found, if we could see it through a sufficiently powerful microscope, to consist of a vast number of molecules, which would be of uniform size and character if the substance were a pure one. No microscope, however, could make such small objects visible, because their size would not permit them to transmit enough light to see them by, and for another reason, which will be referred to presently. Common objects for the microscope are blood-corpuscles and milk-globules. The thickness of the former, or the diameter of the latter, is such that 10,000 could be ranged side by side within the length of an inch. The fifth part of the ten thousandth of an inch covers a very small space indeed in the field of a microscope, and a sphere whose diameter was equal to the fifth part of the thickness of a blood-corpuscle might be taken as an approximation to the smallest object of which a microscope can give us anything like clear vision. A molecule has five thousand times as small a diameter as such a sphere, and 125,000 million molecules would be required to fill it. It will readily be understood, therefore, that there is no possibility of a molecule being seen. Molecules, however, are not necessarily all of the same size. The very heaviest molecule has less than five thousand times the weight of the lightest, and if size is proportionate to weight, their diameter will be sixteen times that of the smaller ones. But even they will show a diameter more than three hundred times, and a surface nearly a hundred thousand times as small as the minimum visible. Such, then, are the individual particles out of which every mass in the world, and, so far as we can judge, every mass in all other astronomical bodies, is made up In the gentlest breezes they blow upon us as a stream of particles finer and lighter than the most impalpable dust. In seas and rivers they flow round us in more closely compacted numbers. In solid substances the molecules are so well anchored in their respective berths, that it requires much force to alter their relative positions sufficiently to change the shape of the mass. But the processes of grinding, dissolving, and fusing show that even in the hardest and densest solids, the rigidity with which the parts are held together is limited, and the molecules can be separated from one another. Thus, in the sixth scale of existence, we see that masses conform to the principle of subdivision which we have seen to characterise matter in all the higher scales. The other principle of relative motion will also be found to hold among the component molecules. And here we begin the study of actions so similar to and so illustrative of those manifested by radium, that the study must be somewhat close and detailed.

Let Fig 17, A, be taken as a diagrammatic plan of the molecular constitution of a solid substance, such as gold or silver, salt or sugar, ice or wax. Here we see the molecules held together by the influence of gravitation, the same energetic force which holds the moon to the earth and the earth to the sun, and which seems to rule the smallest as well as the greatest masses in the universe. The molecules in the solid appear to be so tightly packed that it is impossible for them to move. Yet they are constantly making movements of a certain kind. In the first place, their own construction, which is to be explained afterwards, is such as to admit of elastic changes of shape, so that they can squeeze partly into, and back again out of, the small empty spaces between them. In the second place, though for convenience in illustrating the rigidity of solids they are drawn here as spheres of continuous substance, and closely packed together, it is certain that this representation is incorrect in the former point, and by no means certain that it is correct in the latter. But whatever the conditions in detail of their movements, the molecules are constantly moving. The movement may be of elastic change of shape, as with a rubber ball; of vibration, as in the ends of a tuning fork; of rotation, like that of a top; or when there is larger free space, of translation from place to place, as with a bullet. Or two or more of these kinds of movement may be combined. But what is certain is that all the molecules are constantly in movement, which we will call, for convenience, vibration. The molecular vibrations are exceedingly energetic. In a substance which is incandescent, or white-hot, the molecules vibrate at the rate of 600 million million times a second. For the energy of vibration varies with the heat, being greater when a substance is hotter, less when it is colder. The energy of

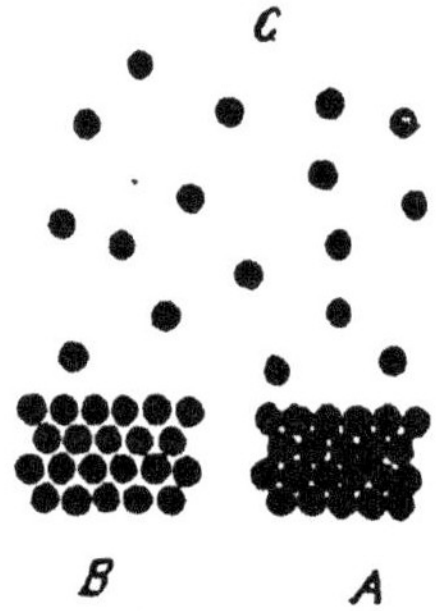

Fig. 17.

molecular vibration is in fact itself the heat of a substance. A consideration of this fact will enable us to understand how heat produces its various effects. How, for instance, does heat cause destruction of substances by combustion or distillation? When a red-hot poker is made to touch a piece of paper, that means that the molecules of the paper are being approached by iron molecules, which are vibrating with a very great deal more than the ordinary energy. They therefore strike much harder blows than usual on anything which they touch, in this case the molecules of paper. Now molecules, as we shall see later, are not single whole masses of continuous material, but are groups of several component parts, often of different elements. The molecules of paper, for instance, are made of carbon, hydrogen, oxygen, and other elements. The vigorously vibrating iron molecules at a red-heat strike them such hard blows as to smash them up and throw apart the constituent portions of the elements which compose them. These are then free, as they cool again, to re-combine with other elements, the carbon of the paper with oxygen from the air forming carbon dioxide, the hydrogen of the paper with oxygen from the air or from the paper, forming water. The paper and some of the air have thus been transformed into other substances. This is the process of destruction by combustion, set up by heat. In destructive distillation the separated parts remain separate, having no oxygen to combine with. How, again, does heat produce pain? Our bodies are provided with nerves running from the seat of sensation in the brain to all parts of the surface. They are capable of transmitting vibratory impulses from any part of the surface to the brain, in much the same way as telegraph wires carry electric impulses from any part of the country to London. In most cases the receipt by the brain of impulses which are unusually vigorous produces the state of consciousness which we call pain. Such excessively vigorous impulses can be set up by cutting, crushing, or chemical corrosion, or by excessive pressure or electric stimulation. So when a hot iron touches the skin, the excessively vigorous vibration of its molecules enables them to deal excessively hard blows to the superficial molecules of the skin. If the heat is not sufficient for the violence to produce actual destruction of the skin molecules, as in the case of burning, still it sets them vibrating as energetically as the iron molecules which keep striking them. They in their turn deal hard blows at the next deeper layer of molecules, which are similarly thrown into a state of excessive vibration. This process of transmission goes on until the excessive vibration, in other words the heat, is conducted right through the skin to the nearest nerve-endings, which then send on excessive impulses along the nerves to the brain, causing the sensation of pain.

This explanation of the nature of heat enables us to understand another of its consequences, the expansion of various substances under its influence. If the adjoining ends of two rails on a railway be examined in winter, it will be found that they are a considerable distance apart. In the heat of summer they will be found to be almost touching, the rails are longer, the iron has expanded under the influence of heat. In the construction of large iron bridges, such as the Forth Bridge, special arrangements have to be made to prevent this alternate expansion and contraction of the metal in hotter and colder seasons from straining and contorting the structure. It is contrived that portions of the iron-work shall be able to slide past one another. How does heat cause this expansion of solid substances? Look again at the diagram showing the molecules in a solid substance, and remember our explanation of the nature of heat. As the substance gets hotter, the molecules vibrate more energetically, that is to say, they strike one another more vigorous blows. In this way they tend more powerfully and more successfully to drive one another farther apart, against the influence of gravitation, which tends to force them together. But the farther they drive one another apart, the more room they take up as a mass. In other words, the substance expands as it gets hotter. Suppose the substance, a piece of wax, for instance, to be made hotter and hotter continually. Its molecules, striking one another with continually increasing violence, and driving one another constantly farther apart, have at last so far overcome the attraction of gravitation, and driven one another so far apart, that, as you see in Fig. 17, B, they now have room to move easily over and past one another, with gentle rolling friction, almost without friction at all. The solid has become a liquid. We now understand the connection between heat and the process of melting or fusion. There are exceptional features in the solidification of water and iron, which appear at first to invalidate this explanation; but we cannot undertake to study all aspects of everything, and as the elucidation of that point will not help us in understanding the action of radium, we will not go into it.

To continue the consideration of the effects of heat. Starting with a substance in the liquid state,—suppose it to be the familiar liquid water,—we will apply further heat to it. The first result is the same as it was in the case of a solid, the expansion of the substance owing to the greater vigour with which the more energetically vibrating molecules strike one another, and drive one another apart, so as to make the whole mass occupy more room. Common instances of this expansion of liquids under the influence of heat are seen in thermometers. Whether the liquid in them be spirit or quicksilver, heating it causes it to expand, and the extra bulk thus produced makes it necessary for the spare quantity to run up the stem of the thermometer, and so indicate the rise of temperature. As the water gets hotter and hotter, another phenomenon, due to the heat, becomes continually more obvious and more important, though it was unobtrusively present even when the substance was in the solid state, as in the case of ice. The blows which the ice molecules receive from one another tend, as we saw, to enable them to overcome the gravitation attraction which holds them together, and so to make them fly away from one another. Some of those on the surface of the ice, driven upwards by blows from the molecules beneath them, and having none above to drive them down, do actually fly off into the air above. What happens to them then will be explained shortly. It is, however, a very small proportion of the surface molecules that spring up in this way from the solid ice. But from the warming water a continually increasing proportion of the surface molecules are driven upwards into the space above. This is natural, as the increasing heat of the water simply means increasingly vigorous vibration of the molecules, so that those on the top receive more powerful blows beneath, while there is no greater resistance above. This process is called evaporation; and it goes on increasing until the vibrations and the blows are so energetic that not only a few of the surface molecules, which happen to receive a specially strong combination of simultaneous blows in the same direction, are driven away from their companions into the space above; but the average number and power of blows received by the majority of molecules throughout the mass is very nearly sufficient to separate them widely from one another. While this is the case, the water which happens at any moment to be touching the hottest part of the vessel, with the hot flame just on the other side, receives that little addition of heat—has its molecules made to vibrate with that little extra vigour—which enables large numbers of them at the same time and in the same neighbourhood to suddenly repel one another so much farther apart that the whole mass occupies over seventeen hundred times its former volume. It is consequently seventeen hundred times lighter than the water which previously occupied that volume, and so it rapidly rises to the top in a succession of bubbles. This is the process of ebullition or boiling, the sudden conversion of a liquid into a steam or vapour, a form of matter of the general nature of gas. A gas is a substance in such an expanded state that its molecules, as in Fig. 17, C, instead of filling or nearly filling the space which the substance occupies, leave a great deal of empty room between adjoining molecules. Through these empty spaces the molecules fly in all directions at high velocity until they strike and rebound from one another or the walls of the vessel which contains the mass of gas. Imagine a room to be pumped empty of air, so that it contained nothing to resist the movement of flying missiles. Imagine a few thousand balls of such perfect elasticity as to have none of their energy lost in bouncing, to be set flying in all directions in such a room, so that they would go on for ever bounding against one another and against the walls of the room. That would represent, on an enormously magnified scale, the structure of a gas, and the action of its molecules, which have had their vibrations, in the liquid or solid state, developed into long flights through space, in the state of a gas. When we speak, however, of long flights through space, it must be remembered that the expression is comparative, having reference to the extreme minuteness of the molecules, and the spaces which separate them in liquids and solids. As a matter of fact, the average distances between two molecules, even when they are flying freely through space, in a gas, is so small that if a cubic inch of air were magnified to the size of a cube of 16 miles each way, the molecules would even then be no more than one-eighth of an inch apart; and the number of collisions that each molecule has with others in one second is nearly five thousand millions.

It thus appears that matter in the sixth scale of existence, a mass, for instance, of one of the substances in the earth, exhibits, as in the higher scales, the two features of subdivison (into molecules), and of relative motion of the constituent molecules. This motion is found to be a vigorous and energetic motion; and before passing to the next scale of existence, it will be well to trace some of its consequences and results which bear most closely upon the explanation of radium. Firstly, then, we will revert to the statement that a molecule can never be seen because it is too small. It was mentioned that there was another hindrance to the seeing of molecules. We are now in a position to understand that even if we could magnify and illuminate the molecules sufficiently for vision, it would still be impossible to see them on account of their rapid vibration, a vibration of many millions of times a second.

The next point to be attended to is the manifestation of a third principle, which we must henceforth join with the other two which we have so far traced through all the scales of existence. This is the tendency of matter to disintegrate. It is entirely a subject of conjecture whether the universe consisted at one time of uniform material evenly distributed through space, and subsequently divided into the masses which formed the stellar and nebular systems. There is, however, good reason to think that the stellar systems have passed through something like a nebular stage from a state of existence in which the matter composing them was still more evenly distributed. The disintegration of the great nebulæ depended upon the contraction of this mass of matter and its division into smaller masses grouped about centres towards which the contraction further proceeded, thus forming the material for the making of separate suns or stars. Similarly in the next stage the sun or star material further contracted and grouped itself not only round its main centre, where now in our solar system is the sun, but laterally also round various subsidiary centres, forming the nuclei of the planets. In this way we have the disintegration of the solar or stellar masses into the component parts of a solar system, the planetary masses. The planetary masses, in their turn, underwent the same process; and as they contracted about secondary centres as well as about the main centre, they disintegrated into the masses out of which satellites were developed. Do single astronomical bodies, such as the satellites or the developed planets, ever disintegrate into the masses of which they are composed? At first sight it appears that the answer must be no. But further consideration makes it doubtful. What are the shooting stars, of which it is easy to see a considerable number at any time of year, if the sky is observed with a little patience pretty frequently? And what is their probable origin? They fly through the atmosphere so swiftly that their friction with the air generates heat enough to make their surfaces incandescent, sometimes enough to make the superficial expansion cause the bursting of the stone, as pebbles burst in a fire. Some of them have been found after their fall upon the earth. Their form and texture suggest that they have cooled from a molten condition which their interior could not have acquired during their brief passage through our atmosphere. Neither does it seem likely that such heat could have been developed by contraction if these masses had been formed independently by the agglomeration of small quantities of nebular material about separate centres, after the manner of single astronomical bodies. Further, they consist of a very limited number of materials, chiefly iron—a selective constitution which implies that they were formed out of parts of larger bodies which had reached a stage of condensation at which the component parts had, to a great extent, undergone separate grouping, according to their density and chemical constitution. These considerations make it very probable that meteors have been formed either by the explosive violence of astronomical bodies in a state of very active eruption, so that they were flung beyond the range of attraction of their parent, or by the violent and destructive collision of two astronomical bodies. In either case the existence and nature of meteors seems to indicate that matter in the fifth scale of existence also, as single astronomical bodies, has exhibited the process of disintegration as well as subdivision and relative movement of parts. In the sixth scale, masses of matter exhibit this property very clearly. We have seen how the vaporisation of water, whether rapidly in boiling, or slowly in drying, is nothing but the disintegration of part of the liquid mass into its constituent molecules, which fly off into the surrounding air in the form of vapour. It is of special interest to us, however, as students of radium, to note that not only liquid masses, but solid ones also, can disintegrate into their molecules. Careful observation shows that ice itself, in a dry air, even though it be constantly below the freezing point, can gradually disappear. It turns slowly into vapour, without melting. That is to say, its surface portions disintegrate into their molecules, which dart off into the air. The late Professor Roberts-Austen discovered that even solid metals exhibit this property. We know that quicksilver, which is a liquid metal, can readily be boiled into vapour, that is, made to disintegrate into its molecules. But Roberts-Austen found that if he placed solid gold and lead in contact, and left them so, some of the molecules of gold would work their way into the substance of the solid lead, where they could, after a time, be detected by appropriate tests. We are familiar with some solid substances which can volatilise quite easily. Such are camphor and corrosive sublimate. These will gradually disappear without melting, by the simple breaking up of the superficial portions into molecules, which dart away into the air. But there is another solid substance, now familiar, though not long ago a scientific rarity, which is more volatile still. This is solid carbonic acid, which is so volatile that, at ordinary pressures, it cannot be melted at all. When ice is heated, its molecules apply the first portions of heat, or extra vibration, which they receive, to getting partly independent of one another in what we call the liquid state, without actual rise of temperature. The heat so employed is called the latent heat of melting. After the water so formed has been heated to 100° C. or 212° F., any further heating is absorbed by the molecules in such a way as to render them quite free from each other's attraction, and enable them to fly through space in the form of a gas, this again without rise of temperature, the heat so employed being called the latent heat of vaporisation. But just as in the case of other substances we have seen that solids can, to a small extent, pass directly into the gaseous state, so this substance, solid carbonic acid, can, at ordinary pressure, altogether omit the intermediate process of liquefaction; and its molecules absorb all the received heat in the form of that vibration-energy which makes them independent of one another as particles of vapour. Solid carbonic acid vaporises so energetically that it may be called a boiling solid. We have now seen that matter through six scales of existence possesses the three characteristics of sub-division, relative movement, and disintegration. A seventh scale has long been known to science, in which these three characteristics still prevail.

We have spoken of the molecules as the smallest portions into which a substance can be divided without destroying its chemical nature. But at the expense of such destruction, we can divide into smaller portions still. That is to say, the molecules themselves can be broken up. For, as in the preceding scales of existence, the molecules are built up of still smaller parts, called atoms. According to the great chemical theory formulated a century ago by Dalton, the smallest existing portions of matter are those atoms which compose the molecule. Their Greek name, atom, indivisible, signifies that the process of subdividing matter must here come to a stop. If we take the smallest existing portion of salt, a molecule of salt, and further subdivide it by chemical processes, we have no longer any salt at all, but an atom of sodium and an atom of chlorine. If we take a molecule of water, and heat it excessively hot, it breaks up into two atoms of hydrogen and one atom of oxygen. Now, just as the molecules of any substance, such as salt, are all exactly alike in weight, chemical affinity, heat capacity, and, so far as we know, in shape and other properties, so Dalton taught that the atoms of one element are all exactly like one another in all these particulars, and that, as the atoms consist of no smaller portions, and are imperishable and unchangeable, we have in them the materials out of which all substances are constructed, just as a house is constructed of bricks or stones. A moment's consideration will show how useful Dalton's theory was in chemical composition and analysis. When once we have clearly grasped the doctrine that every substance is composed of molecules, and that each molecule is made of a definite number of atoms of certain definite elements, the atoms of any one element being all exactly of the same mass, we have reached a very great simplification in the process of combining elements, or of analysing compounds. For it is plain that since no halves, thirds, quarters, or other fractions of atoms exist or can be obtained, substances cannot combine to form molecules in miscellaneous proportions, but only in proportions which are multiples of their atomic weight; so that when once we have obtained from simple combination a knowledge of the atomic weights of some of the elements, that is a great help in ascertaining the proportion in which those elements are present in other combinations. Dalton's theory, accordingly, has been of the greatest possible utility in chemical and physical research during the last century. It may be mentioned, in passing, that most of the elements have their molecules composed of two atoms each, as in the case of hydrogen or oxygen. But a few, such as the vapour of mercury and the inactive gases—argon, krypton, xenon, neon, and helium—have only one atom in each molecule: the atoms can move independently. Compound substances, however, may have a number of atoms in each molecule. Thus a molecule of water contains three atoms; one of sulphuric acid contains seven; and some of the organic substances, those which go to make up the substance of animals and plants, have molecules which are believed to contain in some instances as many as a thousand atoms each. A diagram of such a molecule would look much like one of a stellar system, or very complicated solar system (see Fig. 18). Such a wonderful resemblance is there between nature's methods of construction on the grandest scale and on the most minute.

Fig. 18.

The component parts of molecules possess relative movement also. The atoms keep vigorously vibrating or revolving past or round one another. The energy absorbed in the production of these movements, or developed by their diminution, may account in some measure at least for the heat taken up by some substances and given out by others in the process of combination, as well as for some of the phenomena of latent heat.

And molecules display a capacity for disintegration. We have already seen that they can be broken up into their atoms either by chemical processes or by the application of extreme heat. The disintegration also occurs spontaneously. Nitrous acid in solution slowly breaks up into nitric acid, nitric oxide, and water. This means that molecules of nitrous acid have disintegrated into their constituent atoms, which have then rearranged themselves in fresh ways, forming new molecules of different kinds. But even when no permanent change is produced in the composition of the molecules, they appear in many cases to be constantly undergoing processes of disintegration and reconstruction. In some dances the couples sometimes break up, the partners separating and for a time dancing with other partners, the original pairs being perhaps reconstituted afterwards by a breaking up of the new couples and a fresh coming together of the former partners. So it appears that even when a substance like water remains unchanged as a whole, its molecules are continually being disintegrated into their atoms, which then unite with other or the same atoms to make new molecules exactly like the old. Thus there is a constant interchange going on, in the course of which the atoms from one disintegrating molecule are temporarily free before they unite with the same or other atoms to form new molecules of the same kind as before. It is of this temporary liberation of atoms that the electric current takes advantage, in the process called electrolysis of water, to permanently separate the constituent elements. For the atoms, when independent, become capable in that state of carrying charges of electricity, and since each kind of atom has to carry the kind of electricity which it is able to carry, whether positive or negative, in the direction in which that kind of electricity is flowing, the electric current is able in that way to exercise a guiding influence on the movements of the atoms, so as to direct one kind to one pole, and the other kind in the opposite direction. In Fig. 19, suppose the upper row to represent diagrammatically molecules of water. It is seen that each of them consists of one atom of oxygen and two of hydrogen. Some of these molecules break up, and the atoms composing them are temporarily free. If an electric current be passing through the water, it guides the free

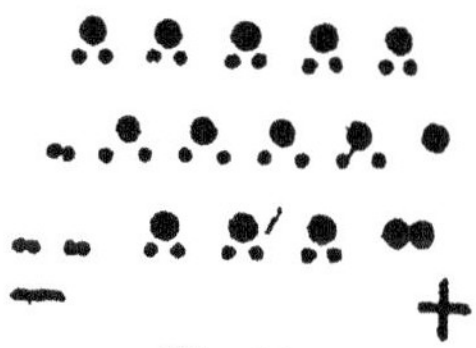

Fig. 19.

atoms of oxygen towards the positive pole, and the free atoms of hydrogen towards the negative pole. Eventually, as shown in the lower line, we have at the positive pole an accumulation of atoms of oxygen which have in their neighbourhood no free atoms of hydrogen with which they can recombine. Consequently they combine with one another, two and two, forming molecules of oxygen gas, which is given off at that pole. Similarly, molecules of hydrogen gas are formed and collected at the negative pole. This is how, in the process of electrolysis, the electric current is able to separate water into its component parts.

We have now seen how matter in the seventh scale of existence, as molecules, repeats for us the manner of construction, and the kind of action which we have found in the higher stages—subdivision, relative movement, and disintegration. But here, according to the doctrine of Dalton, the process of analysis must stop. The atoms being atoms, indivisible, imperishable, the smallest existing portions of matter, it was impossible to pursue the study of the structure and behaviour of matter through any lower scales of existence than the seventh and eighth—the molecules and their atoms. And in this belief science has, for the most part, stood fast and prosecuted its work of fruitful discovery for just over a century, since Dalton enunciated this theory. This is the point at which radium has played an important part in helping to revolutionise the scientific teaching of the last hundred years. How it has done this, we are now fairly well in a position to understand. The preceding explanation of the structure and actions of the matter which makes up the substantial universe has been somewhat prolonged, and has not appeared, during its progress, to be immediately connected with the study of radium. But it is presumed that this book is being read by those who, although they did not begin it with a familiar knowledge of the teaching of modern physical and chemical science, do, in spite of that, desire not merely to know the bare facts as to what radium does, but to understand its actions, as seen from an inside point of view; to learn how it comes to do what it can do, and why it was surprising that such things should be done, and to examine whether its apparent refusal to be bound by the ordinary laws of nature, as previously understood, disappears on closer examination, and is replaced by a deeper harmony between those laws and its actions as special and previously unstudied cases of their application. Such an intelligent appreciation of the facts concerning radium would be impossible to those who are not students of science, without at least as full an acquaintance with the operations of nature in these departments as they have been able to make from the foregoing account.

VI.

Radium and a Revolution in Science.

For some time now a number of thinkers have been inclined to believe that the doctrine of elements and atoms, as recently held by the great majority, was not the last word of science on these subjects. Chemistry had already shown that the thousands of substances by which we are surrounded can be resolved into fewer than a hundred simple substances, called elements. Was it not probable that the same process would be repeated, and the elements themselves shown to be really complex products derived from fewer and simpler radicals? The remarkable gradation of properties displayed by the elements, when arranged on the system of Newlands and Mendeléef, irresistibly suggests a community of origin reducing their differences to mere questions of degree. Some such scheme as that suggested by Crookes, whereby the various elements, with their graduated qualities, were progressively differentiated by a process of development out of some few simple materials, or possibly one only, seemed the most probable view to take of the origin and relations of the elements. This implied that the atom itself should be a complex structure built out of smaller particles; and such a view had been for some time growing among the less conservative of scientific men. The phenomena of the Crookes' tube seemed to require some such explanation, and the view was already widely held that the stream of strangely luminous particles which in such a tube rush from the cathode terminal, exciting Rontgen rays where they strike, and capable themselves of being deflected by a magnet, are not atoms of gas similar to the small quantity which was left in the tube, but consist of matter in an ultra-gaseous state, radiant matter, as Crookes called it; that is, particles smaller than atoms, into which the atoms have, in fact, been broken up under the influence of the electric current. It was a natural speculation, indulged in by some, that matter in this ultra-gaseous state might be so controlled as to have its particles recombined in fresh arrangements, forming other elements than those originally present, and thus imitating the supposed action of nature in the original formation of the elements. There have not been wanting men who have believed that they had accomplished this change, which, of course, is nothing less than the famous transmutation of the elements. Dr Emmens, in fact, made a public announcement, naturally from New York, that he had accomplished the task of turning silver into gold, and went so far as to send across the Atlantic to this country, for examination, some specimens of the product of his process. Investigation, however, did not give satisfactory results. Only part of his silver was converted into gold, and, on the other hand, it was discovered that there was gold present in the silver with which he started. Such experiences naturally confirmed in their conservative views those who had no faith in the newer explanations and the promised developments, who still believed in the atomicity of the atom and the elemental individuality of the elements. It was in this position of affairs that the study of radium compelled an extension of our views of natural operations, which amounted to a new theory devised to explain the remarkable phenomena exhibited by it and by kindred substances. Rutherford and Soddy, of the M'Gill University, Montreal, devised in 1902, and published in 1903, an explanation of these phenomena, which consisted in the view that some of the atoms of radium are constantly disintegrating into smaller portions of matter, and that the energy continually developed by radium is due to the liberation of the energy previously locked up in the form of intra-atomic vibration of the corpuscles of which the atoms consist. Here we have a scientific revolution indeed. For a hundred years Dalton's great and most fruitful theory had held an almost unchallenged position. But one portion of it is now flatly contradicted. The atoms are no longer accepted as imperishable, indivisible, the smallest existing things. The atoms themselves illustrate once more the method of construction seen in stellar systems and solar systems, in masses and molecules; as the diagram of a radium atom in Fig. 20 suggests very roughly, they are constructed of matter in a ninth scale of existence, its component corpuscles being the smallest portions of substance yet known to science. Such a revolution was at first unwelcome to many minds. They clung to less disconcerting, though less adequate, explanations, and rejected Rutherford's proposals as equivalent to an announcement of the suicide of the atoms. Further consideration has shown, however, that it is not only the most adequate explanation yet offered of the phenomena which it was devised to explain, but that it can also explain others which have since attracted attention. This is the experience of theories which are substantially true, and, accordingly, the theory of Rutherford and Soddy rapidly gained adherents, and is now generally accepted.

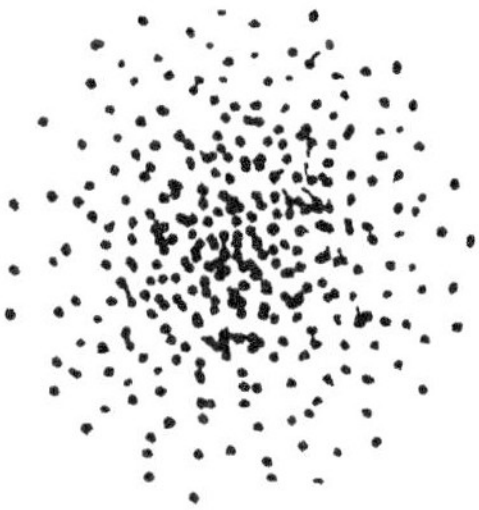

Fig. 20.

VII.

Theory of Radium.

The theory, in greater detail, is as follows:—We are introduced to a ninth scale of existence, the corpuscles or excessively minute portions of matter of which atoms consist. For just as stellar systems contain numerous stars or suns, as masses are built up of molecules, and molecules of atoms, so the atoms themselves are composite structures, containing each a large number of much smaller portions of substance. These are so small that the smallest and lightest atom we know, that of hydrogen, contains from 800 to 1000 of these corpuscles. Radium, however, is a heavy substance, one of the heaviest known, and its atom contains about 200,000 corpuscles. The corpuscles, however, are much smaller than is indicated by saying that their size is the two hundred thousandth part of that of an atom. For they do not nearly fill the space occupied by the atom. They fill as little of the atom space, and are as far apart from one another as would be the case with the same number of the smallest grains of dust floating in the hollow inside of a gigantic football. The skin covering of the football would keep these tiny grains of dust inside, and would preserve the football's size and shape. But if the corpuscles take up so small a part of the atom space, how does the atom, as a whole, maintain its size and shape and its power of resisting external masses? Why do not its tiny corpuscles fall together into a much smaller space? Here we come to the second property of atoms. Like things in the higher scales of existence, they exhibit not only subdivision into corpuscles, but relative movement of the corpuscles. The corpuscles are in a state of enormously energetic movement. Of what nature the movements are, whether concentrically circular, or excentric, or rectilinear and vibratory, is not yet known. Professor J. J. Thomson and Lord Kelvin have put forth ingenious suggestions as to the kind of arrangements and movements of its corpuscles which may enable an atom to hold together as an individual, but all that we can say for certain yet about the movements of the corpuscles, is that they are extremely energetic. We have seen something of the enormously energetic vibrations of molecules, which we know as heat, and the almost countless number of collisions which each molecule of a gas experiences in a second. Yet such tremendous activities as these are rest and quiet in comparison with the far more vigorous and rapid movements of the corpuscles within an atom. This consideration will help us to understand how it is that the corpuscles, which fill so small a part of the space occupied by an atom, can nevertheless give to its surface the quality of resistance and substantial solidity. For so great is the rapidity of the movement of corpuscles that, small as is the portion of surface occupied by any of them, they occupy in a very short time so many different positions that there is no part of the atom-surface which is without the presence of a corpuscle for more than an infinitesimally brief space of time. Thus adjacent bodies will encounter resistance from an atom at whatever part of its surface they approach, and it will possess in some respects the properties of a solid body. It is in this way that we get a practically even pressure all over the inside surface of a balloon or a boiler, though in reality the pressure is not evenly applied all over, but is the result of millions of molecules of gas or steam bombarding the inner surface at points which, though infinitesimally close together, are, nevertheless, distinct and separate points.

The forces which keep the corpuscles within the boundaries of their own atoms, preserving the atomic size and constitution, are, as we have said, yet unexplained. They may be comparable to those which keep the planets, satellites, and meteoric streams to their positions within the solar system, or they may be, and probably are, a combination of much more complex influences and arrangements. But they depend for their operation upon the maintenance of an equilibrium, the destruction of which involves the break-up of the atom. This equilibrium sometimes is destroyed, with the consequent disruption of the atom. The equilibrium of the corpuscular movements may be destroyed by the forcible interference of electric vibrations. Thus in a Crookes' tube with an electric current passing through it, we get the atoms of ordinary gases disintegrated into the material of the cathode rays. But in some substances, such as radium, uranium, thorium, polonium, and actinium, the corpuscular equilibrium is destroyed either by the external violence of inter-atomic collisions, or by the accumulating irregularities of corpuscular vibrations; so that in these cases we have the spontaneous disintegration of some atoms constantly taking place. When the corpuscular movements have lost their equilibrium, some of the corpuscles fly out from the atom, as molecules do from a boiling liquid when the equilibrium between the molecular vibrations and the inter-molecular attraction is destroyed. When some of the corpuscles have thus darted out of an atom, the remainder may settle down into fresh arrangements, and a new equilibrium of new corpuscular movements be established in the diminished atom. The new equilibrium also may prove to be only temporary, not less liable to interruption than the former. This appears to be the case with radium and its congeners. Then further disintegration of the atom follows, and may be again repeated. And while some atoms are in one stage of disintegration, others will be in another stage; so that we get a number of different developments going on simultaneously.

The result is that, in the case of radium, we have the following manifestations.

Radium gives out in the first place a material which is called the α rays. This material consists of small particles which are intermediate in size between atoms of radium and corpuscles. The size of the particles is such that, if the material of the radium atom is homogeneous, they are groups of corpuscles holding together as individual bodies or atoms, and containing each about 1800 corpuscles. The issuing α rays convey by far the greater part of the energy that is developed by radium. This is readily understood when we know that they leave the atom with a velocity of more than 15,000 miles per second. To realise what a tremendous energy this involves, we must remember that the energy of a moving body is proportional not to its velocity, but to the square of its velocity. Of three boys, suppose the second throws a stone twice as swiftly and the third three times as swiftly as the first; then the second stone strikes not twice, but four times as hard as the first; the third stone strikes not three times but nine times as hard as the first One of our most vivid conceptions of force is that of a flying bullet. If a bullet strikes with a speed of half a mile per second, the α particle shot from a radium atom strikes with a speed more than thirty thousand times as great, and therefore its striking energy, weight for weight, is more than 900 million times as great. A great part of this striking energy appears as heat. The α particle is a comparatively large mass, about twice as heavy as a hydrogen atom. It is therefore easily stopped by a thin layer of solid material. It cannot get through a thin piece of metal, or even a sheet of paper. But when it strikes its energy is converted into heat, just as that of a striking bullet is converted into heat both of the bullet and of the object which it strikes. Now, most of the radium atoms are placed internally, being surrounded by other radium atoms. Consequently most of the flying α particles strike adjoining portions of radium and develop heat within the substance of the radium. This, then, is how it comes to pass that radium is constantly giving out heat and yet never gets cool. The heat is merely another manifestation of the energy which was previously manifested in the rapid flight of the α particle. But whence was that energy derived? From the energetic vibrations of the corpuscles within the atom, before its disintegration. This helps us to realise how very vigorously the corpuscles do vibrate. When a fly-wheel or a boiler bursts, the speed with which the broken parts fly out depends upon the energy with which in the one case the fly-wheel was revolving, in the other case the molecules of steam were vibrating against the inner surface of the boiler. Similarly, it is on the energy of movement of the corpuscles within the atom that the enormous speed of their external flight depends. It is not surprising, then, that when this energy of vibration is converted, on the disintegration of the atom, into rapid flight, and the rapid flight into concussion heat, this heat should be very great in comparison with the small mass of material. The heat thus developed by radium in one hour is sufficient to heat its own weight of water from freezing point to boiling point. A quantity sufficient to fill the smallest salt-spoon, if poured into a glass tube, would develop so much heat as to melt the glass. If all the heat ultimately derivable from this small quantity of radium could be converted into work, as our steam-engines convert into work part of the heat derived from the burning of coal, the total heat from the small salt-spoonful of radium would produce work enough to lift a weight of five hundred tons a mile high, or to drive a one-horse-power engine through a working year. The greatest quantity of heat energy obtainable by chemical means is that got from the combustion of hydrogen in oxygen. The total heat spontaneously given off by a definite weight of radium would amount, before it was used up, to thirty thousand times as much as that obtainable by burning an equal weight of hydrogen.

The practical world at once inquires—Is there any means or any prospect of controlling and applying these relatively enormous quantities of heat energy to useful human purposes? The answer at present is negative. The disintegration of the radium atoms goes on at a definite rate without our interference, or the possibility of our interference, either in starting or stopping it. Heating the substance in no way accelerates the development of energy, cooling it causes no diminution. Even chemical combination and decomposition make no difference: bromide, chloride, or nitrate of radium, thorium, or uranium goes on developing its special forms of energy in strict proportion to the quantity of the pure element present. It is true that when some of the intermediate products have accumulated for some time, these products can be removed by chemical processes, with an apparent reduction of the output of energy in the remaining material. But the reduction is only apparent. The separated part and the residual parts go on developing energy each at the same rate as before, though now in different places. So that the only way yet discovered of influencing the output of energy by radium is simply to increase the activity of the material by purifying it, that is, by collecting a greater quantity of it in a small space. This being the case at present, what is the outlook for the future? To judge of this, we must bear in mind that machines are of two kinds—transformers of power and liberators of power. A clock only transforms part of the power which we ourselves have put into it in winding up the weight or the spring. Similarly, a hydraulic lift only transforms part of the power which we put into the other end of the apparatus by pressing upon the water supply. On the other hand, a sluice and water-wheel do not transform any continuous or intermittent work provided by us, but liberate the latent gravitative energy stored in the water of a lake at a higher altitude. And the work done by a steam-engine is not our shovelling and mining work transformed: it is the latent chemical combining energy stored in coal and oxygen, liberated in the furnace before transformation in the boiler and cylinder. By the purely transforming machines we merely obtain some convenience; from the liberating machines we obtain large additional quantities of work. Now we are already able, in the Crookes' tube, to transform electrical energy into some of the special energies manifested by radium—the cathode rays, with the heating effects and Röntgen rays obtainable from them. So far this is very encouraging, since it shows that some of the radium effects are obtainable by artificial means, from other kinds of matter. There is good reason to hope that, with the progress of knowledge and invention, we may learn how to use electrical or other means, merely as a stimulus to disturb the equilibrium of the corpuscles which compose the atoms of common substances, thus causing the disintegration of the atoms and the liberation of the vast stores of energy which they contain locked up inside them in the form of orbital motions or vibrations. When this is accomplished, the human race will be able to obtain, from some perhaps common and economical material, many thousands of times the power that we get at present out of a corresponding weight of coal.

The α rays have a comparatively feeble influence upon the photographic plate, which is so strongly affected by some of the other activities of radium. They are very energetic, however, in the ionization of air. This means the breaking up of molecules of air into atoms or portions of atoms, which are capable of carrying an electric charge, so that whereas ordinary dry air is not a conductor of electricity, ionized air is. When it is remembered that the shaking of steam molecules by heat vibrations can shatter them into their component atoms, hydrogen and oxygen, it does not seem surprising that particles flying at the enormous speed of the α rays, and striking molecules of air, should be able by violence to break them up. At the same time, these collisions stop or deflect or diminish the speed of the greater part of the flying particles, so that the α rays are stopped by a few centimetres, an inch, of air as completely as by a thin sheet of metal.

The particles of the α rays carry a positive charge of electricity. They tend, normally, to pursue a straight course, but are deflected from it under the influence of a strong magnetic field.

It would be out of place in a work like this to attempt to explain how all the results of radium investigations were experimentally arrived at, and the details of the apparatus would, for the most part, be unintelligible to non-scientific readers; but it will be interesting here, as one illustration of the refinements of ingenuity and care required in such work, to show a part of the apparatus by which Rutherford learned some of the facts concerning the α rays, especially their liability to deflection by a magnet. In Fig. 21 we see an air-tight box ABCD, in which is fitted an insulated electroscope EF, of which E is the gold-leaf. GHKL is a box opening into the bottom of the larger box, and having some radium salt spread on the bottom HK. The α rays shot out by the radium, passing into the upper box, ionize the air, and so make it a conductor of electricity. Thus an electric charge given to the electroscope by the movable wire M, and causing the gold-leaf to stand off from the stem, could be carried away by the ionized air, and the electroscope was discharged, the leaf falling down to the stem. The result would have been obscured, however, by the fact that radium sends out other rays which can ionize air. These were neutralised by the following means. Inside the front and back of the lower box were cut perpendicular slots, down which thin strips of metal were made to slide. The rays could pass up the slits between these strips. But the lower box was placed between the poles of an electro-magnet, and the magnetic field easily deflected the other rays, so that they struck the sides of the strips before reaching the top of the slits. Thus they were prevented from ionizing the air in the upper box, and their action was eliminated. But there were other sources of confusion. Radium gives out an emanation, and the smallest quantity of this, passing into the upper box, would ionize the air there, and so spoil the experiment. To exclude this, a stream of gas was passed into the upper box at D, and allowed to leave the lower box at K. As it passed downwards it swept all the emanation out before it had time to enter the upper box. But instead of using ordinary air for this sweeping-out process, Rutherford used a stream of hydrogen, for two reasons. We have seen that a moderate thickness of air is able to stop the a rays, so with air in the boxes very few a rays would reach the neighbourhood of the electroscope. Hydrogen, being much less dense, has much less stopping effect, and consequently will let more of the a rays pierce to the centre of the upper box. Secondly, the other rays, whose action we do not want in the upper box, cannot set up ionization so strongly in hydrogen as in air, so that using hydrogen helps to exclude their undesirable action. Another source of error would be the diffusion into the upper box, in

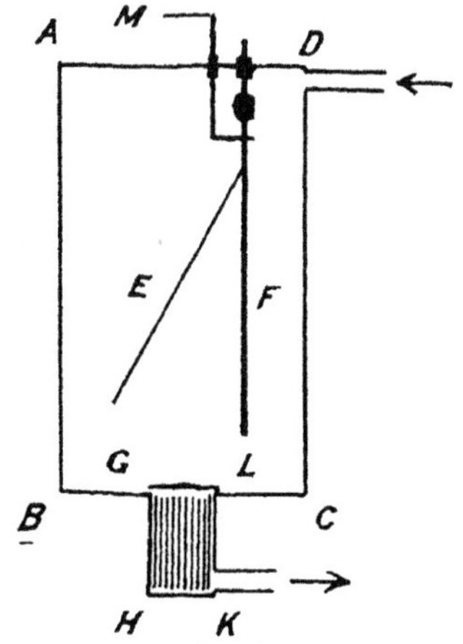

Fig. 21.

spite of the gas current, of ions formed by rays in the lower box before their absorption in the metal strips. To prevent this, a sheet of very thin aluminium foil is laid over the opening between the two boxes at GL′. The metal is so thin as to be porous to hydrogen, and so it allows the stream of gas to pass. But ions, though formed by breaking up molecules, have an electric charge which probably causes them to attract clusters of molecules round them, so that each ion forms actually a mass much larger than a molecule, a mass too large to pass the fine pores in the aluminium foil, so that any ions formed in the lower box are kept there. We now have nothing but the α rays passing from the radium into the upper box, ionizing the gas there, and so displaying its power of discharging the electroscope. If the magnetic field is very powerful, the α rays are deflected a little to the side, so that those of them which are nearest to the strip on that side strike it, and are absorbed before reaching the upper box. Consequently the ionization is weaker, the gas is not so good a conductor of electricity, and the gold-leaf takes longer to fall down after charging. With a stronger magnetic field the deflection of the α rays is greater, they are all caught on the faces of the strips before they can get through the slits, there is no ionization, and no discharge of the electroscope above. It is not yet shown, however, in which direction the deflection takes place, whether they strike on the strip to the right of each slit, or on that to the left. To decide this, the arrangement shown in Fig. 22 is employed. This gives the upper end of a few of the slits and strips, with part of a grating which is laid over them in such a way that its bars overlap one side of each strip, partly closing the upper orifice of the slits, always on the same side. Now a glance at the two slits in the diagram will show that the α rays in their upward course, represented by dotted lines, if they are deflected to the right, must be bent quite across the slit in order to be stopped by striking the face of the strip on the right; whereas, if they are deflected to the left, they will be stopped without being bent more than half across, because the

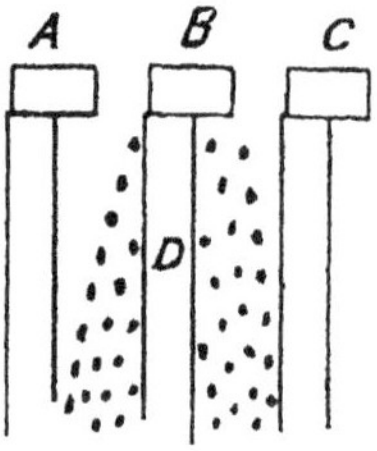

Fig. 22.

projecting bar will stop the last of them, so that they need not bend right across to the strip on the left. Thus, if they are bent to the right, a more powerful magnetic field is required to keep them from reaching the upper box than if they are bent to the left. So by trial and comparison it was ascertained from the direction in which the α rays are deflected in a magnetic field, that they are charged with positive electricity. In this description no reference has been made to earthing of wires, sulphur-bead insulation, windows for microscopic examination, and other complicating details. Enough has been explained, in the essentials of the apparatus, to show that much thought and work are involved in the discovery of many of the facts which we have to be satisfied with simply enumerating. The most interesting and beautiful manifestation of the activity of the α rays is the scintillating light seen in Crookes' spinthariscope, already described. It is open to us to believe that the sparks are caused by the heat involved in the sudden stoppage of the α particles as they strike the screen, in much the same way as we have sparks struck out between flint and steel. Some, however, advocate the view that the sparks are of a similar nature to those seen in sugar or salt crystals crushed in the dark, and that they are due to cleavages set up in the crystals of the zinc sulphide screen by the impact of the α particles. In any case these scintillations are, in the world of minuteness, the most wonderful sight yet seen by the human eye. For they are the work of particles no larger than a molecule of hydrogen. Such a molecule is so small that it would require over two hundred and fifty millions of them side by side to reach an inch. The diameter of one is a thousand times smaller than the smallest thing visible in a microscope, and that is a thousand times smaller than the least thing visible with the unassisted eye. Yet these excessively small objects are as distinctly seen in the scintillations of the spinthariscope as a bullet is when we see the splash that it makes as it strikes the water.

The α particles are not, however, the smallest objects thrown out by radium. They are, as we have seen, groups containing perhaps as many as two thousand corpuscles. But radium shoots out another kind of rays, called β rays. These also are material particles, but much smaller than those which form the α rays. They have the same mass as the corpuscles which form the cathode rays in a Crookes' tube, and weigh about one thousandth of a hydrogen atom. They fly, however, with a much greater velocity, a speed approaching that of light, over ninety thousand miles per second. Their energy, therefore, is, weight for weight, thirty-six times as great as that of the α particles. Owing to their great energy, and still more to their excessively small size, they have a very great penetrating power. Instead of being stopped, like the α rays, by a very thin sheet of metal, they can pass through considerable thicknesses, such as that of a shilling. The general rule, both for α rays and for β rays, is that they pass farthest through the lighter materials, and are most effectually stopped by the denser metals. The heavy metals, however, such as mercury and lead, have a stopping power in excess of their density; so that, though the β rays pass through ordinary materials, such as wood or paper, in much the same way as a bullet through a hedge, they are almost all stopped by a twelfth of an inch of lead, or a fifth of an inch of aluminium. The β particles, like the α particles, are deflected by a magnet, but much more easily; and in this respect also, as well as in mass and approximately in velocity, they resemble the corpuscles of the cathode rays. They carry a negative charge of electricity, which is equal to the charge carried by a hydrogen atom in the electrolysis of water. These β rays are the most active in producing an impression on the photographic plate, and they also ionize the air, though not nearly so powerfully as the α rays. We are now able to understand the action of the radium clock, as it is called, of which mention has already been made (see Fig. 9). The radium in the inner glass tube is constantly firing off, as we have seen, α particles charged with positive electricity, and β particles charged with negative electricity. Most of the β rays, owing to their small size and high velocity, pass right through the glass and get away, carrying the negative electricity with them. The α rays are too large and slow to pass through the glass, and their positive charge remains in the radium cell, which thus becomes positively charged, and induces a repelled positive charge in the leaves of the electroscope. This, however, would be discharged by means of the ions produced in the air by the β and γ rays sent through it by the radium. To prevent this, the radium cell is enclosed in a larger vessel exhausted of air, so that ions are not produced. The leaves of the electroscope thus remain charged, and continue diverging till they touch the sides of the outer vessel, connected to earth, and so get discharged, when they begin the process anew.

VIII.

Radium and the Minute Constitution of Matter.

This appears to be a suitable point, in our investigation of radium, for taking another glance at the lessons which it has helped to teach us as to the structure of material things on what we may call the sub-microscopic scale, and the extremely interesting speculations founded upon this new knowledge. Those who were not, before reading these pages, familiar with the idea of solids, liquids, and gases all being built up of invisibly small molecules and atoms, and even some of those who were, will feel the need of a little more consideration of the atomic structure as explained above. Invited to conceive of the atom of radium as something which, if highly magnified, would be like a balloon without either a gas-bag or a car, but with 200,000 grains of dust whirling about in the space occupied by the former, they may well require some further explanation of the stability of such a system. Why do not the corpuscles, so small and so far apart, fall together as a shapeless heap? Or why do they not fly far apart, with the loss of all atomic individuality? If the corpuscles in an atom are so minute and so far apart, how is it that the corpuscles of adjoining atoms do not get inextricably jumbled up, like handfuls of grain thrown indiscriminately together?

These questions are very difficult indeed to answer. To answer them in detail is in fact beyond any one's power at present, and perhaps will always remain so. But the principles can be indicated on which the results depend, as well as some rough analogies to the manner in which the principles apply. To begin with, it is probable that all depends on the two great principles of inertia and gravitation operating as inexorably among these excessively minute portions of matter as among the mighty masses with which astronomers deal. Under the principle of inertia, every body tends to continue in its state of motion or rest until some other body interferes with it. Under the principle of gravitation, every pair of bodies tend to approach each other with a certain force depending on the distance and the mass of each pair of bodies. The earth is moving at any moment at a certain rate in a direction at right angles to a line between it and the sun. It tends, under the principle of inertia, to continue that motion, which would begin at once to increase its distance from the sun. But, under the principle of gravitation, it keeps approaching the sun, at a speed which just counterbalances the separation due to the inertia movement. Similarly the other planets keep in their position round the sun, and the satellites round the planets; the whole solar system, in fact, maintains its arrangement and its movements, under the two principles of inertia and gravitation. The same principles no doubt explain the constitution of the atom of radium. But inasmuch as it contains something like 200,000 corpuscles, their arrangements and movements are too complex to be calculated in detail as astronomers can calculate those of the heavenly bodies. J. J. Thomson and others have, however, made calculations as to the conditions of stability of various numbers of bodies moving at certain speeds in spherical paths of certain dimensions; and on these calculations have based suggestions as to the construction of atoms. The diagram of Fig. 23 shows the nature of a possible arrangement possessing the necessary equilibrium. The view is supposed to be a sectional one, obtained by dividing an atom into two. Looking at the cut face, we see that the corpuscles are arranged in concentric shells, each moving in its own zone round the centre, as planets move in their orbits round the sun. In the diagram they are shown closer together than is proportionally correct, for the sake of showing more clearly the concentrically spherical arrangement. The corpuscles being so numerous, are liable, no doubt, to disturbances from collisions with

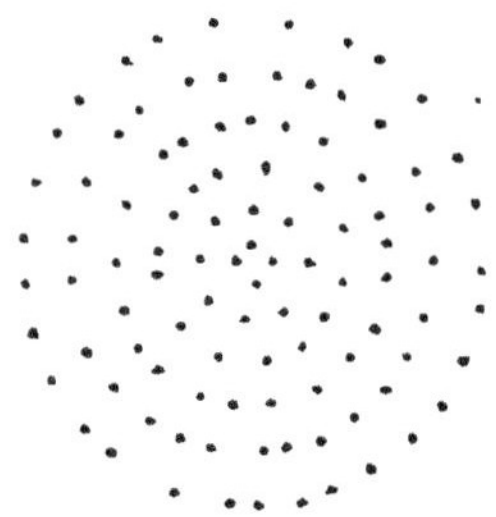

Fig. 23.

corpuscles in their own or in adjoining zones, or from collisions between adjoining atoms. As the comets are members of the solar system whose revolution has been disturbed and made irregular, sometimes to the extent of leaving the system altogether, so we may have irregularities set up in the movements of the corpuscles. Some may pass out of their own zone into another. Some may be driven out of the outer zone, and then fall back into it. Some may fly off so forcibly as to leave the atom altogether and dart into neighbouring atoms. Thus there may be a perpetual interchange going on between the zones and even between the atoms, as some corpuscles wander from one to another. So long as this interchange is fairly balanced, the atoms, as wholes, may maintain their existence unaffected; as we have seen that a liquid and its vapour can keep their proportions while there is a constant interchange of molecules between them, and that molecules in a liquid or gas can maintain their chemical nature in spite of the constant interchange of atoms going on between the molecules. The atoms, however, with their great numbers of corpuscles, are much more unstable bodies than a solar system, and some of them gain or lose so many corpuscles as to have their equilibrium destroyed, with the result that disintegration of the atom begins, some of the corpuscles flying clear of the atomic system, either singly or in groups, while the remainder re-arrange themselves, as an atom of a new substance, which has a very unstable equilibrium, and is consequently liable to undergo further disintegration very quickly. It would be natural to look for the most unstable equilibrium in the most complicated systems, the atoms containing the greatest numbers of corpuscles, that is, the heaviest atoms. And in fact it is radium, thorium, uranium, substances with the highest atomic weight, which display the most active disintegration of atoms, that is to say, the greatest radio-activity.

IX.

Radium and Errors of Speculation.

The corpuscles, then, of which atoms are composed—nearly 1000 to the hydrogen atom, about 200,000 to the radium atom, are the smallest material things known to man. But we must be careful not to confuse the limits of our knowledge with the limits of actual fact. When Fahrenheit had obtained a very low temperature by the mixture of ice and salt, he concluded that there was, and could be, no lower temperature, and accordingly he took it as the zero of the temperature scale which we still use and know as the Fahrenheit scale. Subsequent research has discovered natural temperatures 90 degrees lower, and artificial cold can be produced 425 degrees below Fahrenheit's zero. Again, when Dalton, a hundred years ago, made his discovery that all matter is composed of atoms, he concluded that these were the smallest existing or possible portions of matter, and Clerk-Maxwell called the molecules—using the word, no doubt, somewhat carelessly, with reference to the qualities of their atoms—"the foundation-stones of the material universe," which "remain unbroken and unworn." Seeing that this description of atoms has been much questioned of recent years, and that it is one of the most interesting points in connection with the study of radium, that it has clinched the arguments against such a view, it would be surprising if those who are quoting the properties of radium to illustrate Dalton's mistake, and are teaching the doctrine of atomic subdivision into corpuscles, should repeat concerning their corpuscles Dalton's erroneous presumption concerning his atoms. Yet this is exactly what is now being done. Some of the teachers of atomic disintegration are already calling the corpuscles of which the atoms are composed the ultimate particles of matter, the foundation-stuff of the universe, the urstoff; and though some of these phrases have a more German twang than Clerk-Maxwell's "foundation-stones," the suggestion they convey is not less rash and imprudent. It is quite possible that further research may show the corpuscles themselves to be complex bodies, systems built up of still smaller portions of matter. It is indeed highly probable that this will eventually be found to be true. One of the probabilities in favour of this view is that such a constitution of matter would lend itself to an intelligible explanation of the principle of gravitation, which is still a desideratum. It may be remembered that this principle was invoked to explain the constitution both of atoms and of solar systems; but it would be satisfactory to have gravitation itself explained: and with some modifications of Le Sage's system, this could be done on the hypothesis of further divisibility of the corpuscles. For those who are not acquainted with it, the following may serve as a brief account of Le Sage's theory of gravitation. The universe is filled with something —say the ether—like a very light gas, whose particles are in a state of rapid movement in all directions, so as to beat with

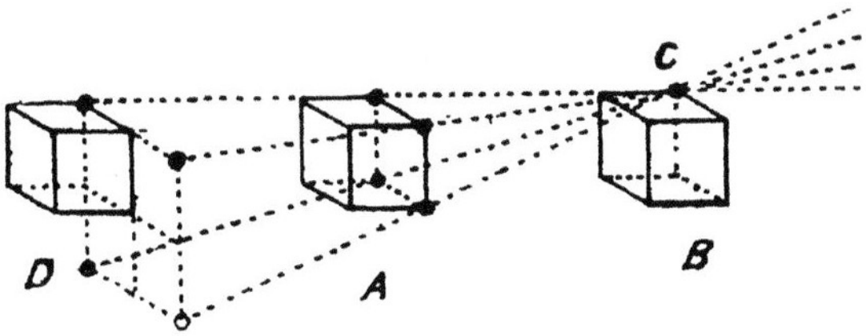

Fig. 24

incessantly repeated blows upon every part of an object of ordinary matter. If an object had one of its sides shielded from these blows, the blows on the opposite side, being unopposed, would drive it in the direction of their own movement. Two objects placed near together shield each other's adjacent sides from many of these blows, and are consequently driven towards each other with a force equal to the difference between the force of the blows on their protected sides, and on the sides opposite. This difference would be inversely proportionate to the square of the distance between the objects. In Fig. 24, consider A as a body of ordinary matter placed in the midst of Le Sage's ethereal medium. It is pounded on all sides by missiles flying from every direction. Of those which pass in all directions through a point at C, a certain number proceed to the nearest side—the right side in the figure—of the body A, and along with all the other missiles which strike it on the right from other directions, they tend to drive A towards the left. But this tendency is counteracted by the effect of other missiles striking it on the left, and tending to drive it to the right. Now suppose another similar body to be placed at B. The part of B which occupies the point through which the missiles were shooting from towards C, will now stop them, and the right side of A will receive so much less battering. Similarly the missiles which formerly struck A by passing through all the other points now occupied by B, will be stopped by B, and A's right side will be saved all their blows. But this will leave all the blows from the opposite directions on A's left unbalanced, and so A will be driven to the right. Similarly, of all the missiles striking A on the left, those which, in the absence of A, would have proceeded to strike B on the left, are prevented by A's presence from doing so, and consequently the similar blows on B's right, from the opposite direction, are unbalanced and drive it to the left. Thus the proximity of A and B causes each of them to be driven towards the other. Hence, that which looks like attraction, which we commonly call the attraction of gravitation, is really, on this theory, nothing but a disguised form of repulsion. This explanation has the great merit of being intelligible, and of having analogies, moreover, on a grosser scale. In the common pump we have something which looks like the attraction, suction, or drawing of water. It is in reality nothing of the kind; but, as is probably the case with all other apparent attractions throughout the whole of nature, it is simply the result of disguised repulsions of different intensities, a yielding where the push is weakest, in consequence of stronger pushes on the other side. The pump works in this way. The weight of the air presses with equal force all over the surface of the water in the cistern, except where it is kept off by what we call the suction-pipe. From the inside of the suction-pipe the air pressure is removed by lifting up the plunger, and so the greater pressure on the rest of the water-surface forces some up the pipe. The water is forced up from beneath, not drawn up from above. The plunger pushes up the air above it, and the rest of the air pushes up the water after the plunger. This will help us to understand how, on Le Sage's theory, what looks like attraction can be merely a special effect of repulsions. His theory also explains how attraction comes to vary inversely, as Newton found, with the square of the distance. Look again at Fig. 24, and consider the diverging pencil of missiles which pass through the point C to strike the body at A. If the body be removed from A to D so that the distance DC is twice as great as AC, the pencil of missiles, when it reaches D, will have diverged in every direction twice as far as at A, and so its cross-section area at D will be four times as great as at A. The missiles in the pencil diverging from the point C, or from any other point in B, will be spread over an area four times as large at D as at A, and so the side of the body at D will be saved by interposing a body at B from only one-fourth of the blows through C from which it was saved at A. And so with the missiles through all the other points in a body at B. Thus B can save D from only a quarter of the repulsion from which it saved A. That is to say, a smaller quantity of repulsion on its left is unbalanced on its right. The excess on the left is only one quarter of what it was before. In other words, the attraction is one-fourth when the distance is doubled. But Le Sage's theory presents two great difficulties. If ordinary matter is being so constantly bombarded by ethereal particles, their surfaces should be constantly raised in temperature. Again, on his theory, we have no explanation of the fact that the force of attraction varies with the masses of the bodies. If a body were once solid enough to stop the flying particles, it would stop no more by being of double the thickness, and yet we know that in the latter case its gravitating force would be doubled.

Would these difficulties be overcome by accepting the theory of atomic subdivision into corpuscles, and supposing further that the corpuscles in turn be built up of a number of still smaller particles, which might themselves be the ultimate granules, or another intermediate structure on the way to the ultimate granules? Let us begin by supposing space occupied, not too thickly, by infinitesimally small particles of material substance flying, at a very high velocity, indiscriminately in all directions, and frequently colliding with one another. Call this material the ether. A number of the flying granules here and there come into sufficiently close approximation to protect one another from a small proportion of the collision-blows which would otherwise strike them on those sides which they turn towards one another. Consequently, they are repelled more strongly on the other sides, that is, driven towards one another, or, as we call it, attracted. Their rapid movements and collisions among one another tend to drive them apart. In some positions an equilibrium is established between the external repulsions inwards and the internal repulsions outwards: the aggregated granules settle down into something comparable to the orbital motions of the planets, and so we have systems formed which may be atomic corpuscles, or may be merely individual groups within the corpuscles. We will suppose them to be corpuscles. Some of them may include greater numbers of granules, and some smaller, thus lending themselves to the production of different qualities in the more complex systems formed out of them. Some of them are more stable systems, and others less so. Individual granules pass occasionally from one corpuscle to form part of another, or to mingle once more with the freely flying granules which still constitute the ether. The formation of corpuscles goes on until, as their numbers increase, and the density of the ether diminishes by their loss, there is an equilibrium established between the absorption of granules in new corpuscles and the liberation of granules by the occasional disintegration of corpuscles. These new bodies, the corpuscles, complex groups of granules, are capable of more complex movements than those of the separate granules. They can undergo compression and expansion, elongation and flattening, rotation and vibration; some of which, singly or in combination, form the rudimentary constituents of those special agitations which we know as electricity and magnetism, a new form of energy or movement being thus developed in the new arrangement of matter or substance. The corpuscles, as wholes, suffer the same repulsion-blows from the flying ethereal granules, which we have already seen to result in holding together a number of individuals which have come into each other's neighbourhood. Thus a number of corpuscles are kept together or attracted in more complex systems called atoms, whose nature varies with the number, and probably with the mass and structure of the constituent corpuscles. These more complex bodies, the atoms, are capable of more complex movements, including chemical energy, transformable into and out of electrical and other forms of energy. Atoms again, under the influence of gravitation, electricity, and chemical affinity, combine into molecules: in couples for the most part, in case of simple or elementary substances, but in larger groups in compound substances; the molecules of some organic substances containing about one thousand atoms. In the molecules we are able to observe that special form of vibration-energy which we know as sensible heat and light, transformable into and out of electrical and chemical energy. Lastly, the molecules combine into masses, which, according to the number, size, and constitution of their atoms, and the physical conditions, may be metal or non-metal, crystal or colloid, oily or watery, solid, liquid, or gaseous.

How does this view of the structure of matter, as a progressive condensation of ethereal granules, help to explain the gravitation energy of masses? Will not the objections to Le Sage's explanation still apply? Will not the blows of the granules be received in the superficial layers of masses, and their energy there converted into heat? In that case, as before, the deeper layers would not count, and attraction would not be proportional to mass. The objection will not hold if the bulk of the granules be very small and their distance from each other comparatively very great, not only where they are flying freely through space, but also where they are somewhat more closely collected in the corpuscles. The flying granules would then, in large proportion, be able to pass not merely between, but right through a large number of corpuscles one behind another, so as to reach the more deeply buried ones. Millions of bodies like our earth could fly right through the solar system without touching a planet, or satellite, or the sun. Suppose these bodies to be of the size of peas, but their distance and number as at present. Then millions of peas could fly not only through one system, but through millions of systems one behind another in a row, and only a small proportion of the flying peas ever strike one of those in the systems being penetrated. If the size of the granules were in some such proportion to that of the corpuscles, the corpuscles themselves being widely separated in the atoms, then the granules could pass through the atoms and through the corpuscles themselves in layers uncountably deep so freely that in flying right through the earth only a very small proportion of the ethereal granules would strike others and be stopped. The matter might be illustrated in another way.

Suppose the granules which go to make up the corpuscles, atoms, molecules, and masses of the earth to be so excessively small that the whole mass of the earth might be crushed flat into a plane passing through its centre and not exceeding a granule's diameter in thickness. They would form, in fact, a single layer in that plane. Suppose them to be so excessively small that even in the thickest part about the centre they occupy only a small part of the plane area, leaving clear spaces relatively so large that of all the ethereal granules flying at right angles to the plane, by far the largest proportion would get through, and only a number relatively small would touch any of the granules in the plane. Then only the same proportion would touch in passing in that direction through the earth as it is. It is on the stopping of these that gravitation, according to this theory, depends. Note that though the greater proportion of the flying ethereal granules get through without touching, yet their number is so great that every granule in the earth gets struck many millions of times in a second. As to the production of heat in the superficial layers of substances by the ethereal bombardment, —the corpuscles situated near the right side of a mass will be to a certain extent protected from bombardment on the left by the mass on the left; the corpuscles situated on the left will be protected to the same extent on the right; those in the middle will be protected to half that extent on each side. Thus there will be the same mean quantity of ethereal hammering experienced all through a mass. It is true that the corpuscles in small masses, isolated at great distances, will be less protected, and so suffer more vigorous bombardment than those in large masses. This will cause the granules of these corpuscles to be driven slightly nearer together before the internal repulsions balance the external ones; in other words, the corpuscles will be slightly smaller. But this need have no connection with heat effects. Sensible heat, as we know it, is a vibration of more complicated structures, the molecules, and even latent heat probably does not affect movements deeper that those of atoms. The principle of infinitesimal smallness of the ethereal granules, in comparison with the mean distance between them, is necessary in another way. To get gravitation effects, the flying ethereal granules must be able to maintain their rectilinear direction, either in themselves, or by transmission to other granules, through distances at least as great as those through which gravitation acts without apparent diminution. Otherwise we should have the same conditions as in ordinary fluids, the liquids and gases we know, through which pressure in any direction is transmitted equally in all directions. This would make it impossible to have repulsion at a distance in one limited direction. So we could not have a differentiated repulsion in any definite direction. That is to say, we could have no attraction. Now the reason why ordinary fluids transmit pressures equally in all directions, is because the mean free path of the molecules is so short that, before one of them has traversed a measurable distance, it has encountered thousands of collisions and been deflected into thousands of different directions. The result is a mean pressure distributed equally in all directions.

Now gravitation, on the above hypothesis, depends on granular repulsions maintaining a constant direction over the distance which separates two mutually attracting bodies. This distance is in some cases astronomical, and so the mean free path of ethereal granules, between successive collisions, must be comparable with astronomical distances. That is another reason for the distances between the granules being enormous in comparison with their size. The distance is not, however, infinite, or we should not get the universal pressure effects of actual experience. The distances not being infinite, a granule will, when it has travelled far enough, have encountered so many collisions as to be sensibly, and ultimately quite, diverted from its original direction. For such enormous distances, then, the granules act as a fluid: the ether, in fact, is a fluid, though astronomical distances are too small to demonstrate its fluid activities. The result is that for these greater universe distances, gravitation no longer applies. We know that it applies, as here, through the solar system distances for which astronomers can calculate and compare its actual and theoretical effects. We can judge from the motions of some of the so-called fixed stars, and from the shape of nebular and stellar systems, including our own, the Milky Way, that gravitation applies over stellar distances, though we cannot say whether with the same force as here. Possibly attraction reaches across the spaces between stellar systems, but, if matter be widely enough extended, there must be distances at which, on the above hypothesis of gravitation, two masses are no longer able to attract one another directly, though they will both be attracted towards each other by intervening masses. It will be interesting in the future to see whether it will be possible to obtain any evidence of the commencement of a weakening of gravitation within the range of astronomical space. In that case, the rate of diminution over such distances will be something progressively higher than the inverse squares. If such progressive diminution begins within the limits of the solar system, it is so small that it has not yet been detected.

One more point needs consideration. If the distances between the ethereal granules are so great relatively to their size, how is it that they are able, by their blows, to exercise a universal and apparently continuous pressure? The answer must be that their velocity of movement compensates for their smallness and rarity. Imagine a large cubical room, measuring 83 feet 4 inches every way. The floor and the ceiling would each contain a million square inches of surface. Imagine a single ball bounding between the floor and ceiling, and making the double journey a million times a second. The blows would be numerous enough, if evenly distributed over the ceiling and floor, to make one per second on every square inch. Let the ball fly a million times as swiftly. Then there will be blows enough for a million per second on every square inch, and, though the ball were as small and light as a pea, it would be a very heavy ceiling that would not be lifted by the pressure of so many blows at such enormous speed. Increase the speed a million times again, letting the blows be distributed evenly all over the surfaces; and not a pin point in the floor or ceiling would escape receiving a million blows per second from a single ball of the smallest visible size. We need nothing, therefore, but sufficient velocity to compensate for smallness and rarity of the ethereal granules, and give us all the effects of gravitational attraction.

We thus see how speculations, partially inaugurated before, but forced upon our closer attention, and more completely developed, in consequence of the necessity of explaining the actions of radium, lead to a possible explanation of the hitherto mysterious force of gravity and the development of the elements; thereby resolving all physical and chemical phenomena into products of two original and simple things —matter or substance, and motion or energy; both of them being transformable, but both also persistent.

X.

A Strange Paradox

So new to our experience, and so paradoxical to our habits of thought, have been the facts concerning radium and the theories based upon them, that some minds have become intoxicated by intellectual excitement; and, stimulated by the lust of novelty and paradox, have proceeded to such lengths in administering shocks to the astonished as to propound views which are not merely beyond the limits of what has been thought possible by men of science, but are, in fact, in direct opposition to the possibilities of consistent and logical conception. In listening, therefore, to new revelations professedly founded upon the theories of radio-activity, it will be well for the uninitiated to put their minds into an attitude of examination. If the new theory be only novel, strange, and astonishing, it may be accepted provisionally, so long as it has the approval of the majority of scientific men. Thus the perpetual output of heat by radium, though beyond previous experience, was not, as explained by disintegration, in conflict with any well-established law of heat. The disintegration itself, though without a parallel in previous chemical experience, was not in opposition to any well-established chemical principle. But if a new theory appears to be irreconcilable with common-sense and consistent thought, as well as subversive of the broadest generalisations which have hitherto had the unanimous support of all men of science, it had better be regarded with suspicion till the whole scientific world have had plenty of time to examine it fully and carefully, and have, without exception, adopted it.

XI.

THE ELECTRONIC THEORY OF MATTER.

Such a theory has already been placed before the world in connection with the new views of the minute structure of matter to which we have been led by the study of radio-active substances. We have seen that atoms are made up of very much smaller bodies, called corpuscles. The smallest atom, that of hydrogen, contains about one thousand of these corpuscles, which are the smallest portions of material substance yet known and measured. We have seen also that each corpuscle carries a charge of negative electricity. The charge amounts to a three thousand millionth part of an electrostatic unit, and is the smallest charge of electricity yet known and measured. The fact that the smallest detected charge of electricity is always associated with the smallest detected quantity of matter, is a connection of an impressive and suggestive kind, and some have come to the conclusion that the nature of the connection is nothing less than actual identity. The corpuscles, they tell us, are charges of electricity. Matter, they say, is made of electricity. Those who believe this seem inclined to adopt the word electron as the name of the corpuscle, and also of its electric charge, regarding these as one and the same. The view they hold is therefore well called the electronic theory of matter. This view, if established, would involve a revolution in scientific thought, compared with which the disintegration of atoms is commonplace; and it must be examined with the care due to its own importance and to that of its advocates. As the space at our disposal here for such an examination must be very limited, we will limit our consideration to some writings by two prominent men whom we may fairly take as representatives of the propounders and expounders of this view: Professor J. J. Thomson, in his Yale Lectures on "Electricity and Matter"; and Sir Oliver Lodge in his Oxford Lecture on "Modern Views on Matter," and his recent article in *Harper's Magazine* on the "Electric Theory of Matter." To do them justice, both these physicists admit that the electronic theory is not proved, either mathematically or experimentally. Thomson says: "Part of the mass of a charged sphere is due to its charge. I shall later on have to bring before you considerations which show that it is not impossible that the whole mass of a body may arise in this way." Lodge says: "Admitting that I am going to promulgate a speculative hypothesis—" and "The formation of an atom of matter out of electricity is a new idea, and has as yet no experimental justification."

On what, then, is it based? Lodge begins his argument by saying: "An electric charge possesses the most fundamental and characteristic property of matter, viz., mass or inertia." Here at the outset is a confusion which, if allowed to pass, would vitiate any conclusion based upon it. Mass is here identified with inertia. If we are using words in their ordinary scientific sense, and there is no warning that it is otherwise here, mass means quantity of matter. Balfour Stewart, Everett, Ganot, and ordinary text-books so define it. Inertia means resistance to change of motion, the work required for starting or stopping, quickening or slowing. It is true that there is a connection between mass and inertia. A heavier ball is harder to throw or to stop when thrown. To put it more technically:—mass, or quantity of matter, is usually ascertained by weighing. But weight is merely the force with which the earth attracts, and this varies with our position on its surface. To get an absolute test of mass which would be independent of position, we may measure the force required to move or stop a body at a certain speed. A four-ounce ball thrown at a speed of 20 feet a second may be stopped by a lever compressing a spring. The inertia of the ball's movement, its resistance to stopping, contains, and is measured by the energy or work of compressing the spring. If the speed is reduced to 10 feet a second, a sixteen-ounce ball will be found necessary for the same compression of the spring. With a speed of 40 feet a second, a one-ounce ball will have mass enough. In either case the compression of the spring would leave it in a position to give out energy enough to overcome the inertia of the ball again and start it back at the same speed at which it arrived, except for friction losses. The inertia then depends quite as much on the speed as on the mass; and it would be quite as legitimate to say that inertia is speed as to say that it is mass. It is, of course, neither.

It is the power tending to resist a certain speed of increased or diminished movement in a certain mass. Note further, that in the above example, while the inertia of the ball has remained exactly the same, compressing the spring in each case to the same extent, the mass has changed between 1 oz., 4 oz., and 16 oz. It is clearly impossible, therefore, to identify mass with inertia. With such a confusion at the beginning of the argument, every step will want watching, to see that some misleading deduction is not based upon it. Dependent on this error is that of describing inertia as "the most fundamental and characteristic property of matter." A more fundamental property of matter is its persistence, since this property does not necessarily involve anything but matter itself; whereas inertia inevitably implies motion, and so is no more characteristic of matter than it is of motion. We need not then accept as correct the remainder of Lodge's sentence, "so that if any one were to speak of a milligramme or an ounce or a ton of electricity, though he would certainly be speaking inconveniently, he might not necessarily be speaking erroneously." May we also hope to be able in time to comfort and cheer ourselves during winter with heat supplied by the gallon, and light by the yard?

Those who would maintain that electricity is matter, must of course prove that it is not a form of energy, as we have long been accustomed to think. Lodge roundly states it as part of the teaching of the disciples of Clerk-Maxwell, which he apparently accepts. "Electricity is not a form of energy, any more than water is a form of energy." If we look out upon the world around us, we see and touch many different things having various shapes, sizes, colours, and other peculiarities. They are found on close study to be transformable to a great extent into and out of one another. These we call substances or matter. Our widest generalisation concerning it asserts its persistence, and we have such faith in this doctrine that whenever matter of a certain kind is transformed into something of different qualities, we have confidence that this new thing is matter too. Besides matter, we observe also changes of position of various portions of matter. These we call motion, or, with a view to some of its relations, energy. Studying motion, we find that it too is capable, as matter is, of many transformations. Thus the motion of a billiard ball may be transformed into the motion of two others, each at an angle to that of the first. The downward motion of a pendulum weight is transformed into horizontal movement as it reaches the bottom of its swing. The falling movement of water is transformed into the revolution of the water-wheel and the grindstones. The reciprocating movements of the piston are transformed into the thousand different motions of looms and other machinery. Further, the quantity of movement, its energy or inertia, as measured by the mass of the moving body and the velocity of its motion, remains always unchanged, or is transformed into equal quantities of energy in other phenomena. This is the doctrine of the conservation of energy, which corresponds to that of the persistence of matter. Nowhere do we find motion disappearing, except to be replaced by other equivalent motion. Nowhere do we find it created, except out of other motions which are ceasing. The movement of a train, when steam is turned off, and it runs on into the station, is transformed into friction-heat in the brakes, wheels, axles, bearings, and rails, and in the air. The moving energy of a hammer-head is transformed into the heat-energy of itself, of the anvil, and of the piece of metal which it strikes. The flying energy or inertia of a bullet is transformed into the heat of the bullet and the target after it has struck. Heat is the vibration of the molecules of which masses consist; and this molecular vibration, the heat energy, can be reconverted into visible movements of masses, as we see in guns, steam-engines, and gas-engines. So we learn that generally, transformations of energy take place not only between the visible motions of gross matter, but between them and the invisible motion of molecular, atomic, and corpuscular matter, and between the latter among themselves. Thus we have heat transformed into the chemical dissociations of water or barium binoxide. We have the chemical energy of oxygen and carbon or hydrogen, of water and alcohol or sulphuric acid, transformed into heat. Heat is constantly transformed into light. Light itself exercises a push upon gross matter, and the force of this push has recently been measured. But in all those cases it is matter that is transformed into other forms of matter, or energy into other forms of energy, and no one has yet seen a grain of matter disappear by conversion into energy, or the concentrated energy of an

Atlantic liner produce as much as an ounce of matter without a full ounce of matter being previously supplied to it. Now, what is the place of electricity in this system of transformations? Is it transformed matter or energy? When a cell has been producing electric current for some time, the original chemical reagents are still present in full weight, though in partly altered form. But their chemical energy is not all present. That is what has been converted into electricity. When an electrical machine or a dynamo is used to charge a Leyden jar or supply electric current for a tramway, there is no change in the quantity of material substance with which operations were started, it is the mechanical energy driving the machinery that has been converted into electricity. When a piece of hot iron is brought near a thermo-electric couple, a current of electricity is produced in the wires completing the circuit. There is no loss of substance in the metals forming the couple or the wires. It is the heat energy of the iron that has partly disappeared, some of it being converted into the electric current. On the other hand, when electricity disappears, it is energy, not matter, into which it is converted. Electric current in wires becomes heat and light and chemical decomposition. Electricity is transformed into magnetic force and into the energy of mechanical movement. But no one ever saw it transformed into any kind of matter. When matter appears in association with electrical action, it is never additional; it has always been present before in another form. We have ample reason, then, for saying that electricity, since it is transformable into and out of energy only, is itself a form of energy, not matter. We shall require a good deal of proof, therefore, before we accept with Lodge the conception of "an atom of matter composed wholly of electricity," or his statement that "matter is composed of electricity and of nothing else." What are the proofs offered? First, there is the statement already quoted that an electric charge possesses inertia. When we have eliminated the confusion between mass and inertia, this tends in no way to identify the electric charge with matter. For inertia, as we have seen, is as closely associated with motion or energy as with matter or mass. The motion of a flying bullet has inertia—it tends to continue unaltered till it is interfered with. But that does not justify us in calling the bullet's motion a kind of matter.

Heat is another form of energy which has inertia. Heat is the motion of molecular vibration, rotation, or other agitation. This motion tends to go on till interfered with. If we protect the heat of hot water as completely as possible from the interference of conduction or radiation by enclosing the water in a silvered vacuum vessel, the heat motion of the molecules will go on for a long time. And when it disappears, it will, like the motion of the bullet, be merely transferred to other places, or converted into other forms of energy. That is to say, heat energy has inertia. All forms of energy have inertia. The fact, then, that electricity has it too is no reason for saying that electricity is not energy. Much importance is attached to the fact that the same electric charge is found in the corpuscles of the β rays from radium, in the corpuscles of the cathode rays, in the ions of ionized gases and of liquids in electrolysis, whether the latter be the small hydrogen ion or the larger chlorine, sodium, etc. The identity of charge in bodies of such different mass, together with the fact that the charge is the smallest quantity of electricity observed, has suggested the idea that a quantity of electricity is composed of a multitude of little bits of electricity, the charge on a corpuscle being one of these bits, an atom of electricity, "which thus steps on to the stage," so Lodge says, "as the fundamental and really atomic substance." Surely the words atom and atomic might have put speculation on its guard at this juncture, when we have just discovered that the atom which has figured on the boards for a hundred years in the rôle of the indivisible minimum can no longer be considered able to play the part. No one knows how soon the atom of electricity will in its *turn* have to step off the stage again, and without the applause which Dalton's atom has earned by a hundred years of useful service. The fact that a quantity is the smallest quantity yet observed is no proof that it is the smallest quantity existing. The history of the atom shows that. And the equality of the charge in different masses, is no evidence that this charge is the least possible, or that electricity is in its nature atomic. What it proves is that the quantity of electrical energy which is developed in the separated individual corpuscles is also sufficient to give direction to separated atoms, and what this suggests is an intimate relation between the nature of electrical and chemical energy, and between the structure of molecules and of atoms. A somewhat similar property in molecules is that every elementary molecule, whatever may be its weight or mass, absorbs or gives out the same quantity of heat between two temperatures. We do not conclude from this that heat energy is something atomic, made up of small bits or quantities, but that the heat vibrations are intimately related to the structure and functions of molecules.

Some of the preceding reasons for believing in the identity of electricity and matter might have been considered equally well in the pages of Thomson, to which we will go for the next argument, which appears to be based on the increase of what is called the apparent mass of electrons. In the following summary and quotation from Thomson, the interests of those who never mastered the first ten pages of Algebra will be considered by substituting the letter n for a slightly less simple expression, and by a little rearrangement of the very simple expressions left. (In the last expression quoted $m+$ has been supplied conjecturally, as probably a printer's omission.)

Thomson considers the case of a sphere whose mass is m, say a corpuscle, which is moving with a velocity v. Its moving energy therefore, or inertia, is, according to the regular formula, $m\frac{v^2}{2}$. But a moving charge of electricity is accompanied by a disturbance of the surrounding ether called a magnetic field, which implies energy. The amount of energy in this case is $n\frac{v^2}{2}$, so that the total amount of kinetic energy associated with the moving charged sphere is $(m+n)\frac{v^2}{2}$. But this would express the kinetic energy of a body, moving with the velocity v, whose mass was $m+n$; so "the energy is the same as if the mass of the sphere were $m+n$ instead of m. Thus, in consequence of the electric charge, the mass of the sphere is measured by $m+n$. This is a very important result, since it shows that part of the mass of a charged sphere is due to its charge. I shall later on have to bring before you considerations which show that it is not impossible that the whole mass of a body may arise in the (this?) way."

No fault can be found with this as far as the end of the statement that the energy is the same as if the mass of the sphere were $m+n$ instead of m. This does not justify the next statement, that in consequence of the electric charge, the mass of the sphere is measured by $m+n$. By Thomson's own hypothesis (his words are, "if m is the mass of the sphere"), the mass of the sphere is measured by m, and the energy of its motion by $m\frac{v^2}{2}$. But he shows that if electrically charged it takes with it when moved a magnetic field involving a further amount of energy measured by $n\frac{v^2}{2}$. The total energy involved in the movement is therefore $(m+n)\frac{v^2}{2}$, and since the additional part of this, $n\frac{v^2}{2}$, is due to the movement of the electric charge, and is equal to the energy that would be involved in the movement of a mass n, we are justified in saying that when a sphere of mass m is moved, firstly plain, and secondly with an electric charge whose magnetic field has a kinctic energy $n\frac{v^2}{2}$, the total energy of movement in the second case is as much greater than in the first, as if the mass of the sphere had been increased from m to $m+n$. But it is not true that the mass of the sphere itself has been increased. This is like saying—The cost of a loaf for lunch is twopence: but I cannot eat it without a slice of cheese, the cost of which is a penny. The whole cost is therefore threepence, which is what it would have been if I had eaten a loaf and a half without any cheese. The price of the bread, therefore, is measured by threepence, and one-third of the cost of the bread is due to the cheese; and it is not impossible that the whole cost of the loaf is due to the cheese. Or again—A gallon of water weighs 10 lbs., but in order to carry it in a can, I must carry the can also, which weighs 2 lbs. The total burden is 12 lbs., which is what it would have been if I could have carried a gallon and one-fifth of water without any can. The weight of the water is therefore measured by 12 lbs.: part of this weight of water is due to the can, and it is not impossible that the whole weight of water may be due to the can.

The fallacy of this is plainly pointed to by Thomson's own words: "The kinetic energy in the sphere is $m\frac{v^2}{2}$; in addition to that we have the energy outside the sphere, which, as we have seen, is $n\frac{v^2}{2}$." And again, after showing that the momentum of the system is increased from mv to $(m+n)v$: "The additional momentum nv is not in the sphere, but in the space surrounding the sphere." The additional energy and momentum being outside the sphere, so is the additional mass.

That there may be a difficulty in following his argument Thomson appears to feel himself, for he tries to enforce it by an illustration, which however unhappily, or happily, makes the fallacy more manifest, since it assures us that we are not misunderstanding him. It is as follows:—

"I should like to illustrate the increase which takes place in the mass of the sphere by some analogies drawn from other branches of physics. The first of these is the case of a sphere moving through a frictionless liquid. When the sphere moves, it sets the fluid around it moving with a velocity proportioned to its own, so that to move the sphere we have not merely to move the substance of the sphere itself, but also the liquid around it; the consequence of this is that the sphere behaves as if its mass were increased by that of a certain volume of the liquid. This volume, as was shown by Green in 1833, is half the volume of the sphere. In the case of a cylinder moving at right angles to its length, its mass is increased by the mass of an equal volume of the liquid. In the case of an elongated body like a cylinder, the amount by which the mass is increased depends upon the direction in which the body is moving, being much smaller when the body moves point foremost than when moving sideways. The mass of such a body depends on the direction in which it is moving." The conclusion that the mass of the immersed sphere is increased when it moves is reached by the fallacy that when a body behaves as if it were a certain thing, it is that thing. The fallacy of this conclusion can be shown both by parallel example and by direct experiment. For simplicity let the sphere to be immersed have a weight of 2 lbs., and a density equal to that of the frictionless liquid. Let the sphere be placed and left at rest in the liquid. Being of the same density as the liquid, it neither floats nor sinks. Outside it had a weight of 2 lbs. In the liquid it behaves as if it had no weight, as if it had no mass. Therefore, according to Thomson's reasoning, it has no mass; and we deduce that spheres immersed in liquids of equal density are entirely deprived of mass. We know, of course, that it has a mass measured by the weight of 2 lbs., just as before, and that the falling energy, due to the action of gravity on this mass, is counterbalanced by the rising energy, due to the displacement of an equal mass of liquid on which also gravity acts. Now cause the 2 lb. sphere to move horizontally. Its motion involves the energy due to the movement of a mass of 2 lbs. But certain quantities of liquid move with it at various velocities, the resultant energy of liquid movement being, according to Green, equal to that of 1 lb. of liquid moving at the same rate as the sphere. There is then a total energy of movement equal to that of a mass of 3 lbs., and Thomson tells us that all this mass is in the sphere, whose mass has been increased from 2 lbs. to 3 lbs. by its movement. Now the attraction of gravity is proportional to mass: therefore the earth pulls the sphere down with half as much energy again as before: it is half as heavy again. But the surrounding liquid which it displaces is no heavier. It is more likely to be lighter, since the additional mass of the sphere must have come out of the liquid. The liquid, therefore, is now no longer able to buoy up the sphere of increased mass, and it sinks. At least, it should sink if Thomson's deduction is correct. But since experimental trial shows that it does not sink, there must be some error in the proof that its mass is increased. The error is in gradually slipping out of the expression "behaves as if its mass were increased" into "the mass is increased." Even the former of these expressions is not quite correct. "Behaves" should be modified by adding "in one respect only, as regards the kinetic energy, which is partly its own and partly that of the surrounding medium, transmitted through it." For in other respects the sphere immersed in liquid does not even behave as if its mass were increased. We have seen that it does not sink. If it is iron, its influence on a magnet is not increased.

If it is glass, its index of refraction is unchanged. If it is wood, its porosity is not diminished. From such a very limited meaning of "behaves as if it were" we should be doubly on our guard against deducing "is." The careless and incorrect use of "mass" does no harm as a shorthand expression, so long as it is used simply for the measurement of certain energies. When the whole question at issue, however, is the exact location of certain masses and energies, the confusion between "is" and "behaves in one respect as if it were" is inadmissible. It creeps in in various other disguises. In parts of the argument it puts on sheep's clothing, in the form of the phrases "apparent mass" and "electrical mass." The "apparent mass" of a sphere is used, as we have seen, to mean the actual mass of the sphere, together with a mass of something else moving with the sphere. This something else is probably the ether in some manifestation. The whole of phenomena are probably summed up correctly in Thomson's declaration "that all mass is mass of the ether, all momentum, momentum of the ether, and all kinetic energy, kinetic energy of the ether." Again, "electrical mass" has no correct meaning at all. Mass is quantity of matter, and can be neither thermal, chemical, nor electrical. The phrase may be otherwise convenient, but its chief result is the confusion produced by denominating in one phrase, as if they were one thing, the actual mass of a corpuscle, and the imaginary mass which, if moving with the corpuscle, would have the kinetic energy developed by some form of ethereal vibrations set up by a moving corpuscle electrically charged. Used with strict exactness, the words "electrical mass" have no more meaning than "narrow colour," "fragrant velocity," or "yellow scent." Presently, however, the word "apparent" or "electrical" is discarded, and the poor corpuscle, like an overloaded donkey, is required to bear as its own the whole burden of its real matter-mass, and the additional hypothetical mass corresponding to certain electro-magnetic energies developed by the motion of the charged corpuscle. The next step is to say that the real mass of the corpuscle is of the same nature as the hypothetical "electrical mass." Then the increased mass, having now all become "electrical mass," is called electricity without more ado, and we have the full-blown electronic theory: matter is not matter, but electricity.

The electronic theory has taken up a good deal of our attention; but not without good reason. It is one of the most remarkable outcomes of the studies which have been so much stimulated by the properties of radium. The number of men who have adopted it, and the confidence with which they themselves hold it, have both been much exaggerated in the public thought. The Prime Minister, in a speech much praised for its philosophic insight, has apparently accepted the view that the new doctrine puts electricity or ether in the place of matter. Preachers, inevitably regarding the electricity which is to take its place as a form of energy, and therefore something of a higher order than matter, are already comforting themselves and their congregations with the idea that the latest tendency of science is less materialistic in its interpretations of nature. Those, however, who have recognised that fashion has an authority, if not so arbitrary, yet quite as real, in scientific thinking, as in clothes, will not hasten to get beneath the roof of the electronic theory, till they see it able to stand without so many fallacies for props.

The true lesson of the Premier's address is the doubt he suggests as to whether the electronic theory will remain fashionable with the scientific world. "Whether the main outlines of the world-picture were destined to survive, or whether in their turn they were to be obliterated by some new drawing on the scientific palimpsest——." And he gives, in reasoning which may be summarised as follows, good grounds for hesitating to put the electronic theory on the same level of certainty as the generally accepted laws of nature. No argument can be considered sound which professes to prove in its conclusion that one of its premisses or postulates is false. All the ascertained laws of nature concerning solids, liquids, gases, masses, molecules, atoms, corpuscles, energy, velocity, heat, light, undulation, radio-activity, and all other phenomena whatever—the laws concerning all have been arrived at by reasoning which has for one of its postulates the duality of motion and matter. This postulate has its origin in the necessities of human thought. Whether we regard phenomena as subjective states of consciousness, or as realities external to ourselves, it is as impossible for the sane mind to believe in motion taking place without there being something which moves, as to believe in the sum of two and two being less than four. If sane people tell us that they believe in such things, they may have sincerely adopted the statements, but they have not clearly realised their meaning. Common-sense and science having accepted it as a postulate that when there is motion there is something moving, that which moves has been by common consent called matter, or measured in relation to other quantities of matter by certain standard units, mass; and motion measured in relation to antecedent or consequent motion by standard units, is called energy. These conceptions have pervaded the whole of scientific reasoning, and any conclusions inconsistent with the premisses necessarily involve fallacies in the reasoning. Some of the fallacies have been pointed out above. The false conclusion is that corpuscles are electricity. Now electricity, as all good physicists have held, is a form of energy convertible into the energy of mass movement, heat energy, light energy, chemical energy, nerve stimulus, and muscle stimulus. It is convertible, also, out of most of these forms of energy, theoretically out of all. Energy itself is motion considered in a definite relation. If, therefore, corpuscles are electricity, matter is motion, which is contrary to one of the postulates on which all scientific studies, including those of corpuscles and electricity, are based. Lodge in one place meets the difficulty by declaring roundly that "electricity is not a form of energy." As he has identified electricity with matter in "the conception of an atom of matter composed wholly of electricity," it, of course, cannot be energy to him. In that case matter, besides its ancient names,—matter, substance, mass,—has got another name—electricity; and we are without a name for those special forms of energy which are manifested in an electrical machine or a Leyden jar, in the dynamo or the tramway wires. We shall have to find a new name for the energy of those special vibrations of matter or substance, which are not the vibrations involved in the energies known to us as heat, light, or chemical action, but convertible into and out of those vibrations—a new name, that is, for the energy hitherto known to us as electricity. But to give to the matter or substance which moves the additional name of electricity, thereby leaving without a name that particular variety of its motion whose energy has long been called electricity, is uncompensated confusion and loss. Thomson meets the difficulty by referring everything, in the ultimate analysis, to the ether. "All mass is mass of the ether, all momentum, momentum of the ether, and all kinetic energy, kinetic energy of the ether." This statement, whose probability amounts almost to certainty, implies two principles, both quite true. Firstly, that all matter, in whatever form or quality, whether mass, molecule, atom, corpuscle, or universe-vibration-medium, is ether. Secondly, that since the kinetic energy of any mass of matter is a different thing from the mass of matter, as the kinetic energy of a bullet is not the bullet itself, so no kinetic energy in any form, whether mass movement, heat, light, chemical energy, or electricity, is identical with ether. The first principle, the identity of matter and ether, involves that a corpuscle is a quantity of ether in some form. The second principle, the duality of ether and kinetic energy, involves that a corpuscle is not electricity. Yet the contrary of this is Thomson's teaching in places. For though, sometimes, in describing the corpuscles of which the material atoms are built up, he says, "The corpuscle carries a definite charge of negative electricity," thus differentiating the corpuscle, as matter, from the electrical energy of which it is the seat, yet elsewhere he says, "I have proposed the name corpuscle for these units of negative electricity." And we have seen that he holds that part or all of a corpuscle is really the kinetic energy of the magnetic field established by a charge of electricity.

The origin of this confusion appears to be as follows: —

Experiment and calculation established the fact that in all the corpuscles obtained, whether in the cathode rays of a Crookes' tube or the β rays of radium and similar substances, the charge of negative electricity is always the same. This might easily be understood on the established system of physical knowledge, by supposing that every corpuscle in a free state was the seat of the same amount of that particular vibration energy which gives us the phenomena of electricity, just as we know that every molecule of simple solids or gases carries between two temperatures the same amount of that vibration energy which we call heat. It was further learned that the same amount of electrical energy characterised varying masses, in free atoms of different substances, whether singly, as in electrolytic solutions, or in groups, as in ionized gases. The same equality of heat energy is similarly found in the molecules of various elementary substances, which differ much in weight or mass: and, in this case, it merely raises the interesting question why varying masses should absorb or give out equal amounts of energy between two temperatures. But in the case of the corpuscles and their electric charge, as this charge is the smallest known, instead of the question suggested being—why varying masses, corpuscles, and atoms in the free state, with or without attendant clusters, should have equal amounts of electrical vibration, the question was asked—Is not electricity a particulate thing, composed of small portions like the molecules of solids, liquids, and gases, the smallest possible portion being the charge of a corpuscle? This idea, and the phrase "carrying a charge," made it necessary to think of each atom conveying its quantity of a quasi-material electricity about with it like a bundle, or in some cases, for there was no disguising the activity of electricity in comparison with ordinary matter, the bundle was supposed to drag its bearer about, as, for instance, to the positive or negative pole of a battery. In the case of the corpuscle, its bundle of electrical energy, when it was in very rapid motion, became so great, that its own real mass was comparatively nothing, and it was possible to look upon the bearer as all burden—the corpuscle was electricity and nothing else. But inconsistencies, once introduced, lead to endless others; and real simplification will be sooner obtained by keeping the principle that no kind of matter, corpuscular or other, can be identical with any kind of energy, such as electricity; the hope being retained that Thomson's words quoted above will in the end be proved true, by our learning how all kinds of matter, including corpuscles, are merely varying arrangements of granules of ether, and all kinds of energy, including electricity, varying modifications of the movements of those granules. Here we will leave the fascinating question of the minute ultimate structure of matter and analysis of energy, one of the most interesting studies which have received so strong a stimulus from the discovery of the properties of radium.

XII.

Other Features of Radium.

Besides the α rays and β rays, radium sends out a third kind called the γ rays (alpha, beta, gamma). The γ rays are not material substance, like the other two, but vibrations of the ether, very similar to those known as Rontgen rays. They are set up by the sudden starting or stopping of corpuscles. These, in the cathode stream of a Crookes' tube, are suddenly stopped by a platinum disc, and this sudden stopping causes the ethereal vibrations known as X-rays. In the case of radium they are caused by the sudden starting of the β particles or corpuscles, which, as we have seen, fly out at a speed comparable with that of light. They have a very high penetrating power, higher than that of most Röntgen rays. Their effect can be observed after passing through an inch of solid lead, or twelve inches of iron. They are, unlike the α and β rays, not deviated in an intense magnetic field. They have the power of acting on a photographic plate, and of ionizing air. They produce luminosity in willemite and platino-cyanide of barium.

One of the most remarkable and interesting features of radium is the fact that it is constantly giving off some substance which passes into the air around it, and is called the emanation. To understand it, we will first look into a very remarkable property of another radio-active element, thorium. If thorium nitrate be dissolved in water and then precipitated by ammonia, it is found, when collected, to have lost pratically all of its radio-activity. On the other hand, the solution is radio-active, and as the water is all evaporated, and the ammonia salts removed by heat, the small residue remaining is as radio-active as the original thorium. So far as this residue is visible it is merely impurity, for its quantity is so small that the loss of thorium is imperceptible. Small as it is, it is a new substance, formed by the disintegration of thorium atoms, and is called thorium X. It is constantly being produced and stored up in the thorium salt, and is then separated by the precipitation with ammonia. Next to the fact that the thorium X possesses all the radio-activity and that the thorium has lost it, the most remarkable point about these substances is that the thorium X at once begins to lose its radio-activity, and in three or four weeks has lost it all, while the thorium, which had completely lost its radio-activity, begins at once to recover it, and in three or four weeks is as strong as ever. The explanation is that thorium has constantly some of its atoms breaking up into another form of matter which we will call thorium X. The atoms of thorium X are stored up in the substance of the thorium among the molecules of the unchanged atoms, until it has accumulated in what is, for radio-active purposes, a considerable quantity, though the change is really so slow that it would be a thousand million years before all the atoms of thorium had been so transformed. But the atoms of thorium X undergo a further disintegration, which is accompanied by the usual evidences of radio-activity in the emission of various rays. Consequently when the thorium X has been removed from the thorium, the latter has practically no radio-activity. But some of the thorium atoms begin at once to turn into thorium X atoms, which undergo further disintegration, with manifestations of radio-activity, and continually accumulating in the thorium, end by making it as radio-active as ever. Meantime the thorium X, an imperceptibly small quantity to begin with, has been getting gradually used up as it disintegrates further, its manifestations of radio-activity getting less and less until they have totally ceased by the time that the thorium has recovered its activity. After thorium has been converted into thorium X, the next step in its disintegration is the production of a material called emanation. Here we may return to radium, which produces an emanation, but without going through a preliminary change: there is no radium X. The emanation then produced by radium is something given off from it which passes into the surrounding air, where it indicates its presence by displaying radio-activity similar to that of radium itself. The air containing it may be removed from the neighbourhood of the radium, being drawn along a tube, or carried in a bottle, and it will still show a capacity for ionizing the air and so enabling it to discharge an electroscope, or for producing phosphorescence on a piece of willemite. Moreover, the material thus conveyed in the air from the radium is deposited on the walls of the tube, the sides of the bottle, or any other solid object exposed to it, especially a cathode surface; so that all these things acquire from it the property of radio-activity. It is absorbed by water exposed to the air containing it, so that this water becomes radio-active. It passes with the air through porous substances. It is expelled from the radium in comparatively large quantities by heat, or by solution of the radium salt in water. In the latter case the water absorbs the emanation, which can be taken up by air bubbled through it. But when the radium has in these ways been deprived of its emanation, it remains for some time without any, and only gradually becomes possessed of increasing quantities, recovering before the end of four days half the quantity that had been removed from it. It is evidently an accumulation, as in the case of thorium X, of something continually produced in small quantities by the disintegration of radium atoms. The emanation itself, by further disintegration, gradually loses its activity, and in less than four days its power has sunk to one half of what it was at its liberation from the radium. To determine the nature of these emanations was no easy task. They could not be made to show any signs of chemical reaction. Either they were some kind of inactive substance, like the gas argon, which will combine with nothing, or the quantity of them was too small for any chemical action to be evident. For indeed the quantity ever obtained is so little that it could never be weighed by the most refined weighing instruments. How, then, could anything be learned of its nature? At this point of the investigation the author has an interesting personal connection with it, for he happens to be the inventor of an apparatus for making liquid air, which is used in various laboratories in this country and abroad. It occurred to Rutherford and Soddy, in their researches, that they might learn something of the nature of radium emanations by exposing them to very low temperatures. Now liquid air provides the lowest temperature readily obtainable; a very low one indeed; its boiling-point being twice as far below the ordinary temperature of the air as the boiling-point of water is above it. With the object of trying the effect of the low temperature of liquid air upon the radium emanations, Professor Cox, of the physical department in their own university (the M'Gill University, Montreal), got the author to show one of his liquefiers at work in London. They soon had one at work at Montreal, and before long the author received a letter informing him of the success that attended the first employment of his machine to make liquid air for the purpose of testing the emanation of radium. The essentials of the apparatus employed in demonstrating the properties of the emanation are shown in Fig. 25, though the appliances by which the exact estimations were made were naturally much more elaborate. Some liquid air is first made and placed in the vessel D, which is then raised until the spiral glass tube above it is immersed in the intensely cold liquid, which makes the spiral also intensely cold. The tap C being closed, and the taps A and B open, a stream of air containing radium emanations is very slowly passed through the tube from A to B. The extreme cold of the glass spiral, surrounded by liquid air, causes the emanations to be liquefied and condensed in the spiral, so that the air coming out at B shows, if tested, that it no longer has the radio-active properties which it possessed when it entered at A. The tap A is now closed, and by means of the tap C connection is opened to the glass tube E, which contains in its upper part a zinc sulphide screen F, and at the bottom a piece of willemite G. Through the tap B connection is now made with a vacuum pump, which extracts the greater part of the air both from the tube E and from the spiral tube. The emanation, however, being liquid at the low temperature, remains in the spiral tube. The tap B is now closed, and the vessel of liquid air D is lowered away from the spiral tube, so that the latter begins to be warmed by the heat of the atmospheric air outside it. When

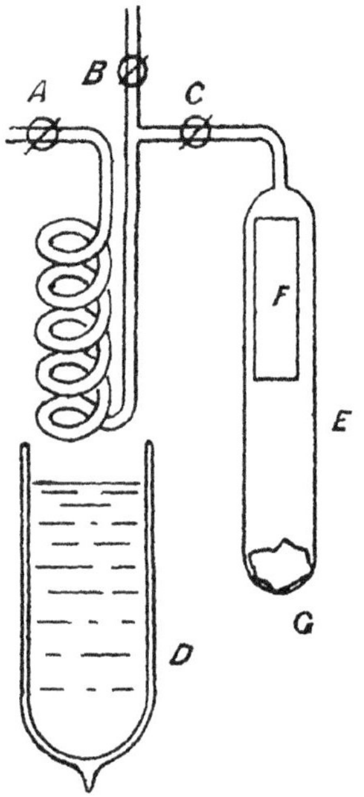

Fig. 25.

it is warm enough for the volatilisation of the liquefied emanation, the emanation, turning into gas, diffuses through the tube towards C. This movement is helped by the fact that the small quantity of air remaining in the spiral tube now expands as the temperature rises, and so helps to carry forward the emanation past C into the larger tube E. Its presence in this tube is indicated by the phosphorescence of the screen F, for this is an effect which we have seen that radium emanation can produce upon zinc sulphide. In the same way, and for the same reason, the piece of willemite G

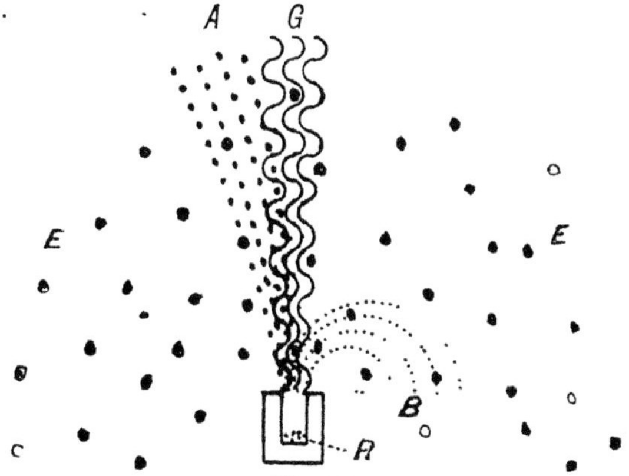

Fig. 26.

A, α rays: positively charged: slightly deflected.
B, β rays: negatively charged: easily deflected.
G, γ rays: ethereal vibrations: not deflected.
E, Emanation particles. R, Radium.

also becomes phosphorescent. If the vessel D be now brought up from beneath over the lower part of the tube E, so that the glass in the neighbourhood of the willemite becomes intensely cold, the emanation is again liquefied by the cold, and condensed upon the glass in that part, so that the greater quantity of the emanation collects there. The more abundant rays from this increased quantity of emanation cause more brilliant phosphorescence in the willemite, while the screen above, being now less actively influenced, becomes less brilliant. When the liquid air is again removed, this end also grows warm again, and at the proper temperature the emanation volatilises once more, and indicates its re-diffusion through the tube by restoring the phosphorescence of the screen. These changes clearly prove that the emanation is a gas condensible at a low temperature somewhere above the boiling-point of liquid air: a gas, probably, of high density, but without chemical activity, as we at present understand it. On more accurate examination, with more elaborate appliances, it was ascertained that the temperature at which radium emanation condenses and volatilises is $-150°$ Centigrade, that is, 270° of frost on the Fahrenheit scale.

The illustration in Fig. 26, modified from Mme Curie's diagram, shows in one view the chief products of radium disintegration, without, however, the confusion which would result from an attempt to show the secondary disintegration of emanation particles. To illustrate the deviability of some of the rays in a magnetic field, the radium is placed in a deep box of lead, so that all the rays have to start in the same direction, moving vertically upward.

XIII.

The Transmutation of Elements.

It has already been shown how radium has given us practically perpetual motion in the constant output of heat, or in Strutt's radium clock, which depends upon a constant production of electricity. It now remains to show that we have also in the action of radium an equivalent of the philosopher's stone. This has been shown in the course of further elucidation of the properties of radium emanation, made, this time, in the old country, and with the assistance in this case also of one of the author's liquefiers, in the employment of Sir William Ramsay at University College, London. It has been mentioned that the radium emanations undergo continued disintegration, until they finally cease to be, or cease to give evidence of their existence by any radio-active processes. The question arises, What has become of them? What is their final form at the end of the process of disintegration? And how do they manifest themselves? When the research concerns such excessively small quantities of material, it might seem hopeless to look for success. It must be remembered that, owing to the great rareness of the material, one-third of a grain of pure radium bromide is a very large quantity, such as extremely few experimenters are ever privileged to handle. Nearly all medical work with radium is performed with a fraction of that mass, or about 10 milligrammes. The emanations produced by such a specimen of radium in twenty-four hours would be a tiny bubble of gas, occupying a volume of about one-fortieth part of the size of the smallest pin-head. It is the final product of this, after it has undergone further disintegration, whose nature has to be examined; and in the experiments conducted on the most lavish scale, only twenty-five times this amount, that is, about half a tiny pin-head of the emanation, was available for examination. Happily the work, in Ramsay's hands, was in those most competent to handle small quantities of gases. In the wonderful work in which he discovered the new inactive gases of the atmosphere, krypton, xenon, and neon, using liquid air as a means of extracting them and purifying them by fractionation, he had been dealing with substances which exist in the air in proportions so small as to be measured by millionths, and in the course of these experiments a volume of gas equal to that of a small pea was frequently a large allowance admitting of lavish application, and comparatively easy spectroscopic analysis. But the energies of radio-activity are so remarkably great, that by its means we can discover with the electroscope the presence of radio-active material, such as radium, in a proportion 150,000 times as small as the smallest distinguishable by means of the spectroscope. So that we might have a tiny quantity of emanation, easily recognisable by its ionizing action in the radio-active state, which nevertheless, when disintegrated into other material, no longer radio-active, would leave a final product so small in volume as to be beyond the reach of the grosser method of spectroscopic examination, which has hitherto passed for the most delicate method of chemical or physical analysis, and still remains so in the case of material which does not give the manifestations known as radio-activity. The refined methods, however, which Ramsay and Travers had developed in their researches on the rare gases of the atmosphere, proved equal to identifying the residual gases from a minute speck of radium emanation. Part of the apparatus employed for this purpose by Ramsay and Soddy, at University College, sketched by the author while he saw it at work, is shown in its actual size in Fig. 27, and forms another good instance of the delicate researches required. The taps H, C, and D are closed; the bulb A, containing radium bromide, is exhausted of air, and the bulb B contains water freed from air. The tap E being closed, and the taps F and G open, all above E and D is exhausted of air by a vacuum pump connected to G, which is then closed. The tap C being opened, water flows down and dissolves the radium in A, liberating the emanation, which is partly dissolved in the water, and producing small quantities of oxygen and hydrogen, since radium has the power of decomposing water. Some helium is liberated also, which has been accumulating in the radium. A vessel of liquid air is raised beneath the U tube K until the latter is immersed and cooled down to the temperature of the liquid air. The tap D being opened, the emanation and other gases partly expand into the upper part of the tube, and the emanation is condensed in the U tube, which is now very cold. Oxygen admitted through H washes out more emanation and helium and hydrogen from the solution, carrying them upwards. The

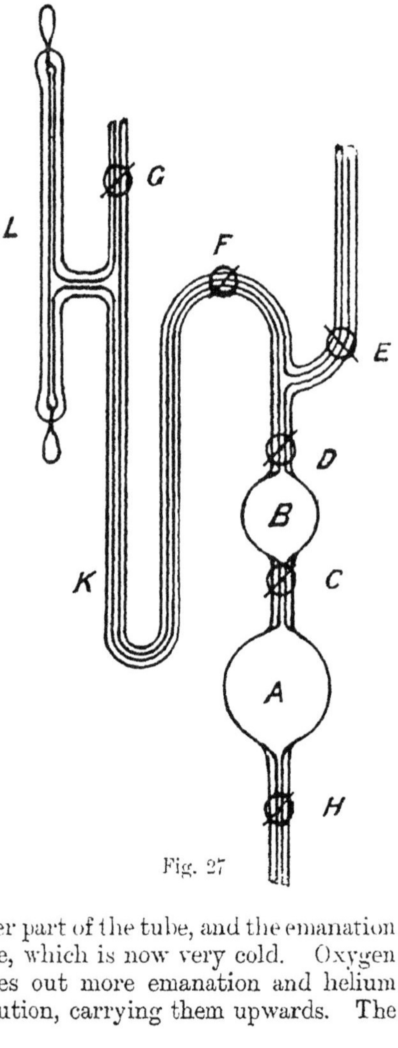

Fig. 27

tap G being now opened, the volatile gases are drawn off by the vacuum pump, but the emanations, being more easily liquefied, are condensed and caught in the U tube. The process of admitting and pumping off oxygen having been repeated, and a good vacuum produced with D closed, G is also closed, and mercury, admitted by opening E, runs up to F. The liquid air being now withdrawn, the emanation condensed in K begins to volatilise as the U tube warms, and part of it expands into L. When the U tube is well warmed, the mercury, being allowed to rise further, sweeps the emanation almost completely into the tube L. This is a spectrum tube with the wire loops for the electric current seen at the ends. It was, as the author saw it, of the same size as represented in the figure, the bore being that of an ordinary pin, and the length not two inches. The pressure of the emanation gas in this tube was very low, only a few millimetres; so that at ordinary pressure it would occupy only the two or three hundredth part of its actual volume, in fact, as we have said, half that of the smallest pin-head. The gas in L was now examined by the spectroscope. At first it gave an unknown spectrum, which Ramsay and Soddy took to be the spectrum peculiar to the emanation. But after three days the well-known D line appeared, and at the end of five days all the lines of a well-defined spectrum. The spectrum was one which it was dramatically very fitting that Ramsay should be privileged to find. Some years after Lord Rayleigh and he had discovered the first known of the inactive gases, argon, Ramsay discovered another in cleveite and other minerals. Astronomers many years before had noticed in the sun's spectrum evidence of a substance not known elsewhere, which they accordingly called helium, from the Greek word for sun. This was the substance which Ramsay discovered as an inactive gas occluded in cleveite; and having prepared a pure specimen, he observed its spectrum, and found it to be as shown at D, Fig. 28, identical with the one long previously obtained from the sun, as at C. The chief lines of the radium spark spectrum are shown at B, and of hydrogen, for comparison, at A. It may be mentioned that the spectrum obtained from the phosphorescent light of radium is identical with the nitrogen spectrum, and merely shows that radium has the power of rendering nitrogen luminous as the electric current does. The spectrum obtained after gradual development by Ramsay and Soddy from the material in the small tube where they had enclosed the radium emanation is shown at E. It is evidently the same spectrum as shown at C and D, the spectrum of helium. The experimenters had no option but to conclude that they had helium in their tube. How had it got there? They had placed in it some radium emanation. This then must, by continued disintegration, have been converted into helium. The emanation itself had been, by previous disintegration, converted out of radium. The conclusion was inevitable, that they had witnessed the transmutation of radium into another, a distinct and separate element.

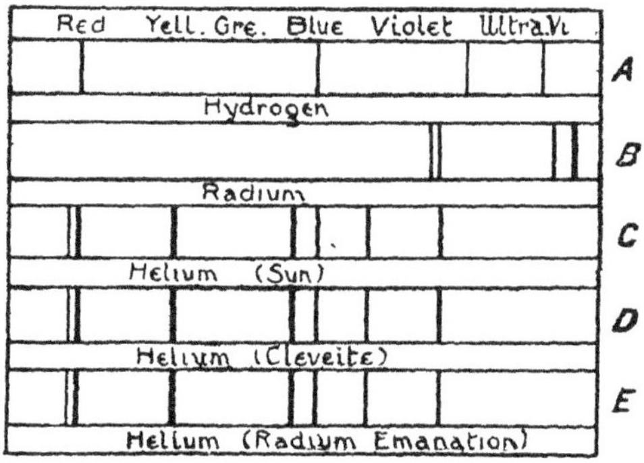

Fig. 28.

Here was at last the realisation of the alchemists' dream. They had for centuries striven to find the means of turning the baser metals into gold. Our object is a slightly different one. We know that if we could turn our paving-stones into gold to-morrow, gold would cease to have for us a great part of the value which it has to-day. But the change of one element into another is, so far as it goes, a justification of those who have held that, underlying the differences between the various elements, there is some simpler and still more elementary material or materials of which all are made. This discovery is the most interesting, and theoretically the most important, in the whole history of science. Practically this transmutation may be the first of many others, made at our will, to suit our own purposes.

XIV.

Future Possibilities.

At present these disintegration changes are entirely beyond our control. We have no hand in starting them; and we have no power to stop them. But it is not necessary to believe that our power will always be so limited. If anything like the progress that the last eighty years have seen in our knowledge and command of electricity be made in the study of radio-activity, we may learn how, by appropriate stimuli, electrical or otherwise, to set up and also to control in common materials the disintegration-change which occurs spontaneously in radium. Something like this can already be done in the Crookes' tube, where the cathode rays are a very close imitation of the β rays of radium, and disintegration of the gas atoms has already taken place. If we learn how to set up such changes on a large scale without unreasonable expenditure of energy, we may be able to produce large quantities of useful elements or compounds out of cheap and ordinary substances which have at present no special usefulness, and which the chemistry of to-day has no means of converting for us.

In the production of power the effects of the study of radium may be still more stupendous. At present our mechanical power is derived, much of it directly, the rest indirectly, from heat. And we have seen that in total output of heat, radium is weight for weight thirty thousand times as productive as the best combustible at our disposal, hydrogen, which, again, is four times as productive as coal; so that one pound of radium would produce, before being used up, more heat than fifty tons of coal. Converting the heat into work in the usual way, what a fruitful source of energy we have in radium! It is not, however, to radium directly that we must look for any possible revolution in the production of power. The material is much too scarce for such purposes; and there is no likelihood of its ever becoming plentiful enough to be of practical use as a substitute for coal or oil. Here, as in the field of chemistry, the possibility of great developments in the future depends on our success in learning, from the study of the minute structure of matter, as revealed by the phenomena of radium and others of a similar nature, how to control the disintegration of atoms, so as to unlock and withdraw the enormous stores of intra-atomic energy which are no doubt present in ordinary and common materials. We shall have to be able to produce the first of a train of events, taking the same place in reference to a series of disintegrations as the striking of a percussion cap holds in relation to the spread of combustion through grains of gunpowder. It will further be necessary to contrive that the disintegration shall take place at such a rate, and in such a manner, as will admit of advantageous employment. There is no reason, however, to think any of these things beyond the bounds of possible accomplishment. The situation differs widely from that of liquid air at the time when it began to arouse wide popular interest. Charlatans then made the wildest promises on behalf of liquid air, confidently asserting that in a few months it would supersede steam as a source of power and ice as a means of refrigeration. Owing to the thermo-dynamical principles involved in the production of liquid air, the fulfilment of these promises was from the first impossible. It is quite otherwise with radium. The material itself will probably never be plentiful. But no one can say confidently that it will not be possible to obtain on a large scale, in common materials, that immense proportional liberation of heat energy which we witness now in radium.

XV.

Radium and the Cure of Disease.

One of the points that have excited the greatest public interest in radium is the fact that it has a powerful action upon animal tissues, definitely curing certain skin diseases, and having, as it appeared, some effect upon some of the conditions which are grouped together under the name of cancer. Radium was at once boomed as a cure for cancer. In this the public have been misled by the press.

In the early days of radium Curie discovered that it injuriously affected the tissues of the body. Having had a quantity in contact with his arm for some time, he found that a bad ulcer afterwards developed there. The sore was very persistent, and took months to heal completely. The most interesting feature was that it also took a long time to appear, and weeks to become fully developed. Others had a similar experience, and it was found that those who continuously worked a good deal with radium were liable to deterioration of the skin of the finger-ends, with cracking and roughness where the parts had been much exposed to the action of radium. These results led physicians and surgeons to try its influence upon various diseases of the superficial tissues. It was soon found that cases of lupus showed immediate amendment under the treatment. The inflammation at once began to subside, shortly cicatrisation set in, and ultimately the sore was replaced by soft and healthy skin. The cure had every appearance of being complete, and though much longer time must elapse before its permanency can be confidently affirmed, the cure is in the meantime one that gives sufferers the greatest satisfaction. For lupus cases the radium treatment is a very valuable addition to the Röntgen rays and Finsen light, which had already been doing such good service upon them. On rodent ulcer also it has a very beneficial effect. The author has seen cases of this disease of a very malignant type, which had shown no improvement under the Röntgen ray treatment, distinctly benefited by treatment with radium. Various skin diseases, such as eczema and psoriasis, have also sometimes derived benefit from the radium rays. There is evidence, from a number of medical men, that superficial nodules of cancer, as they are called, have disappeared under radium treatment, and not returned. The evidence on this head is very positive. On the other hand, some have entirely failed to get any good result in such cases. In view of the widely different characters of growths which are all called cancerous, and of the difficulty of being certain as to the nature of a particular growth, it is not possible to say definitely that cancer has been cured, even when speaking of small superficial nodules. As to deeply-seated, widely-spread cases of long standing, it is evidently hopeless to deal with those by means of radium at present, and such attempts as have been made have resulted in complete failure. Nothing remains beyond the assurance of some that simple cases have been cured, except the possibilities of the future. With such a wonderful agent as radium, which has already effected amazing cures of other diseases, it would be unwise to say there can be no hope of success with cancer. Hitherto the quantity of material we have had at our disposal, and the means we have had of trying it, have both been extremely limited. But in both respects we may look for improvement. Already we hear of American devices for flooding the body with some substances in which radium rays will afterwards excite special activities, tending, of course, to cure diseases. Whatever may be the fate of any individual proposal from America, we must as yet take care not to lend any encouragement to the belief that radium is now a cure for cancer.

It is an interesting question how radium effects a cure in the cases in which it undoubtedly does this. It will help us to understand this if we remember how radium has a remarkable power of producing injuries. And this, in its turn, will be more intelligible if we consider another fact, that radium can impart a very durable activity to things which it comes near. We have seen how the emanation deposits on surfaces something which is itself radio-active. But something with similar powers is deposited in the interior of substances. A quartz crucible which had held some radium for a few days, became radio-active itself. The active material could not be burned or washed off by the severest treatment, and it persisted for months. Some active materials had been deposited within the solid substance, and these went on disintegrating and displaying further activity. So we can understand that the same thing happens in the case of the skin, when exposed to the action of radium. The γ rays, being merely vibrations of the ether, should not have a continued action. The β rays, being the smallest particles known, should not be liable to further disintegration. The α rays, so far as can be judged from their weight and other characteristics, are atoms of helium, final results of the disintegration, final products, as we have seen, of the emanation. But helium is not known to exist in a radio-active condition. The emanation, however, is capable of further disintegration. It is radio-active. It may be, then, that some of the emanation atoms, receiving violent blows from the energetic α and β particles, acquire an impetus which is able to carry them through the superficial epidermis into the deeper tissues. Whatever be the mechanism of the action, it is fairly certain that some particles capable of further disintegration shoot or are shot from the radium into the tissues. There, as they disintegrate, they in their turn discharge tiny β particles, with a velocity of about 100,000 miles per second, and somewhat less tiny α particles, at nearly 20,000 miles per second, into the surrounding cells. Where the missiles strike with such tremendous energy, they naturally produce some damage. In the sparks of the spinthariscope we see with our eyes the effect of the violence. Probably some of the cells so bombarded have their vitality destroyed. But the number being small, no distinctly painful effect is produced. As the process goes on continually, however, and probably gets more intense, each emanation atom giving rise by disintegration to a number of disintegrating particles, the accumulation of the débris of dead cells clogs and poisons the tissues, and sets up the inflammatory decay of tissue which is called an ulcer. This explanation accounts for the slow development of the sore, reaching its worst a long time after the application. It accounts, also, for the persistence of radium sores, since it is only when the disintegration material is getting exhausted that healing action can begin to overtake and repair the destruction. We can understand now how radium can cure diseases, at any rate when they have a microbial origin, as in the case of lupus. Lupus is caused by microbes, practically identical with those of consumption, settling and multiplying in the tissues of the skin. The result of applying radium is that the missiles of disintegration which fatally strike some of our body-cells, fatally strike the microbes also. Our cells and the microbes both suffer together from the radium fusillade. But not with the same results. For the microbes are attacked also by the defence cells and juices of our bodies, specially provided for the purpose. It is only when our defence cells and juices are not strong enough to do their appointed fighting that the invading hosts of microbes triumph, and we fall victims to disease. Where the defence cells and juices are just too weak to guard their territory against invasion, they may be able to prevail if reinforced by the friendly armies of an allied power. That is what happens when radium cures lupus in our bodies. The invading microbes, opposed by our own defensive lines, and bombarded by radium particles at the same time, succumb to everwhelming force. But, it will be objected, the radium bombardment attacks our own cells as well, and so do as much harm as good. The first part of the objection is true, but whether the second be so depends upon the discretion with which the radium is applied. If the supply of disintegration atoms is only moderate, their fusillade will not be able to destroy our cells faster than they can be absorbed and their places filled by multiplication of surrounding cells, assisted by the detergent and nutrient juices supplied by the circulation of the blood. At the same time, the attack may be forcible enough, in combination with our own powers, to destroy the microbes of disease. If this be the explanation, the governing necessity is for caution as to the duration and strength of the application. This corresponds with the facts in practice. While a careful application, judiciously repeated, will produce a satisfactory cure, excessive use of the agent causes very bad sores, for which the radium, instead of the operator's carelessness, sometimes receives the blame. In one case recently the irritation maintained by constant application of Röntgen rays and radium rays, succeeded, as other kinds of irritation sometimes do, in provoking an attack of that scourge, cancer, with fatal results. It would appear that the Röntgen rays had been applied much more than radium. We may compare the curative action of radium to a method which might be adopted for destroying tigers in a piece of jungle, in preference to the plan of actually entering it. Riflemen might be posted round the selected piece of jungle, with directions to keep firing in at random. Some of the bullets would strike and cut down some jungle grasses. But these, if the fusillade had not been too severe, would afterwards grow again from the roots, nourished by the soil and its moisture; whereas any tiger that was hit in a vital place would be done with altogether. So the tigers might be destroyed without permanent injury to the vegetation.

Some hold, however, that the curative power of radium is due to stimulation, as in cases of electrical treatment. We know both that the material rays of radium carry electrical charges, and that the γ rays are ethereal vibrations very closely resembling the Röntgen rays. It may be that these play an important part in some cases; but in a distinctly microbial disease, such as lupus, it is extremely probable that the greater part of the result is due to the mechanical violence of disintegration.

The method of applying radium is to have it mounted in a small case. It looks like a few small grains of salt, and few are fortunate enough to have a quantity which would make a large-sized pin-head. At first it is white, but with age it grows brownish. It may be mentioned here that a glass tube containing radium, or a glass vessel containing a solution of radium, is gradually turned violet or brown in colour. To keep the grains of radium in place, and to protect them from moisture, they are covered with a thin disc of mica or aluminium. This, however, stops some of the β rays and a large proportion of the α rays, thus rendering it less active externally. The radium thus mounted is held close to the part to be treated, either in the hand or in a mechanical holder for a period of from five to ten minutes every few days. If a large surface has to be treated, the radium is exposed to different parts of it successively.

Soddy has proposed to apply radium or thorium in the attempt to cure consumption by the following method, to which several medical men are giving a trial. The radium bromide, or thorium nitrate, is dissolved in water, which receives the emanations. Air is bubbled through the water to draw off the emanations, and the air containing the emanations is inhaled by the patient. The hope is that some of the active material may be deposited within the lungs at the seat of disease, as well as elsewhere, and may, by destroying the microbes, give the patient an opportunity of recovering. In view of the harm radium particles can do if employed in excessive quantities, it is obvious that the greatest care must be used in such work. The physician has to grope, as it were, in the dark, since the effect of his treatment may not appear till much later. Experiments have been made as to the effect upon frogs, mice, and other small animals, of keeping them for a few days in an atmosphere containing radium emanations. They have sickened and died in a few days, sometimes apparently from paralysis, due to the action upon the spinal column and nerves, sometimes from an affection of the lungs which made breathing difficult, and so produced suffocation. This is a forcible caution on the subject of administering radium to human beings. On still lower organisms the presence of radium exercises remarkable effects, such as retarding the development of larvæ.

Other interesting questions suggest themselves in connection with medical work. It has long been noticed that certain cases of goître were cured by allowing the patients to drink no water but fresh rain-water or distilled water, and this though their ordinary well water or tap water could not be found to have any deleterious impurity in it. This has long been a puzzle, but we may have an explanation of it in the facts of radium. For, now that attention has been so carefully directed to the matter, it is found that radium, or some radio-active substance, is present in places where it had not previously been looked for. It is not difficult, as a rule, to tell whether the radio-active substance, whose effects are observed anywhere, is radium or some other. For, whereas some of them have an emanation, which, like that of radium, is capable of imitating some of the effects of radium, and of exciting an activity in other things by leaving on them a deposit which is also radio-active, they differ from one another in the duration of the activity of the emanation and of the secondary activity excited by it. The radium emanation takes $3\frac{3}{4}$ days to lose half its power. The thorium emanation suffers the same proportional loss in one minute. By such tests, conducted with the electroscope, Strutt has ascertained that the Bath waters contain radium. It has been found, also, in the waters of Buxton, Matlock, and other places. At Bath a reddish-brown deposit, in some places solid, in others spongy, accumulates on the surface of pipes or stones exposed to the water; and in this material also Strutt found the presence of radium. There can be little doubt that some, at any rate, of the curative powers of famous waters are due to their radio-active properties. This, however, is not all. Thomson has found evidence of radium in the Cambridge tap water, and an appreciable amount of radio-activity may be detected by the electroscope in most water from wells and springs, even when the quantity of material which gives rise to it is too small to be detected by the most delicate chemical analysis. This may explain the phenomena of goître. Our constitutions differ widely in their susceptibility to different influences.

While many derive benefit from waters distinctly impregnated with radium, it may be that others will suffer from much smaller quantities, and possibly the thyroid gland is the first organ to be affected. Thus it may well be in consequence of the small amount of radio-activity in well and spring water, that cases of goître have developed which are cured by simply confining the patient to rain-water.

XVI.

Radio-Activity in General.

But radium is distributed more widely still. It has been found in the soil of our back-gardens, and in the flour from which our bread is made, in the chalk of the hills, and in the sand of the sea-shore. It pervades the air, and a negatively-charged wire exposed for some hours to the atmosphere is coated with radio-active material, which can be rubbed off with a cloth, or dissolved in acids. The air from the depths of the earth, or from underground passages or cellars, is richer in this property than ordinary air. It is found, too, in the leaves of plants, and, in fact, the difficulty now begins to be to name a place where there is no radio-activity at all.

The truth probably is that, just as radium is an extreme case among the radio-active substances, disintegrating more rapidly than thorium, uranium, polonium, or actinium, so these altogether are merely an extreme case of ordinary matter. There is good reason to think that every element, or almost every element, has a few of its atoms constantly disintegrating; but as the process is much slower than in the case of radium, it escaped notice till the most refined methods of observation were brought to bear. When radio-activity was observed in various metals, it was at first supposed that the manifestation was due to the presence of radium in small quantities as an impurity, but it is now recognised that they have a radio-activity of their own. All the elements in the universe, it appears, are undergoing slow destruction. We cannot point to the hardest or the most solid and say, here is a substance whose particles suffer no loss by degradation. The actual rate of degradation is, however, very slow. In the case of radium, the most energetic, the number of atoms that actually disintegrate in the course of a year is one in about two thousand. Thorium and actinium last much longer. They lose in a year one atom out of two thousand millions. Yet thorium and uranium are distinguished from common substances by their great activity in this respect, and the life of these common substances is therefore immeasurably longer. When we have realised that these phenomena are shared, though in small degree, by common substances, we may discover therein the explanation of many baffling puzzles. It has been an insoluble problem how an almost endless amount of scent-sensation could be produced by an almost infinitesimally small quantity of material. A grain of musk can, without apparent diminution, distribute scenting material through the frequently changed air of a room for years. A dog can track men and animals by smelling the immeasurably and almost inconceivably small amounts of boot or hoof material which they have left on the ground at each step. It may be that the olfactory nerves are sensitive to impressions not only, perhaps not at all, from molecules and atoms, but from the corpuscles into which they disintegrate. It is by no means necessary to suppose, because the corpuscles all have the same electric charge, that they are all of the same kind. They may be of complex and varying structure, as the atoms themselves are. And so they may affect the sense organs in different ways. And the simultaneous presence of different corpuscles in various combinations may produce as many and complex smell-sensations as there are compounds of the elements. So excessively small quantities of material in the radio-active state may be able to produce as definite and appreciable effects upon the consciousness through the olfactory nerves as upon the electroscope by ionization of the air.

A puzzle which appears capable of solution with the assistance of radio-activity is that of the duration of the earth as a geological body. When we dig deep down we find that the earth is one degree Fahrenheit warmer for every 50 feet of descent. Lord Kelvin had calculated that in reaching this condition it could not have occupied more than one hundred million years from the time when it was a molten mass. The interval which this would have left for the deposition and modification of the various strata as we find them was many times too short to satisfy the geologists whose business it is to study those matters. It now appears that the proportion in which we find radio-active material in the earth, if continued to its centre, would provide such an output of heat by disintegration, as would multiply many times the duration of cooling, and allow an ample margin after satisfying the largest demands for millions of years' time made by the geologists. A similar supposition regarding the sun, the supposition that a small fraction of the sun's substance is radium, will enormously prolong the calculated period during which that body will be able to give out heat enough to maintain animal and vegetable life upon the earth. Thus we see that the facts of radium, whose study we began by looking at the constitution of the mighty stellar systems, has taught us truths about the operations of nature in the almost infinitely little, to be reflected back again in the form of knowledge which helps us to a better understanding of the great astronomical bodies whence we get the warmth that maintains our life.

It will be asked, with all this radiation following disintegration, what becomes of the material which is radiated, and what is the fate of the original stock of radium or other material? We have seen that the α rays are probably helium, neither more nor less. We have seen that the emanation ends by producing helium, whatever else it may have produced in intervening stages. Thus helium, which is found in the atmosphere, in waters, and in minerals, is one of the last products of radium. The β rays, or individual corpuscles, have not had their destiny traced. It may be that after their flight, when they meet again in sufficient numbers and under suitable circumstances, they recombine to constitute atoms, perhaps of helium. Failing that, perhaps they wander at large, forming a kind of gas more attenuated and impalpable even than radium, and possibly transmitting special forms of energy whose existence even is not yet suspected. What happens to the radium itself is naturally that it keeps getting less. It is calculated that at its present rate of decay, half of a stock of radium would be dispersed in about 1500 years, and at the end of 10,000 years only about one per cent. would remain. Uranium and thorium change much more slowly, a million times as slowly, so that the time required for reduction to one-half would be 1000 million years, and to come down to one per cent. about 10,000 million years.

But on this estimate that 10,000 years would suffice for the reduction of the quantity of radium to one per cent., how is it that we have any left? The rocks in which it is found have existed for more than a thousand times 10,000 years, and even if they had been full of radium to begin with, there should have been none obtainable now. It is necessary to suppose that from some source the radium is replenished as it wastes, so as to maintain a balance between the production and the loss. In this case the question arises, out of what is it made? Observing that the disintegration process involves a change from larger or heavier and more complex atoms to smaller and simpler ones, and that the only two atoms we know of as heavier than radium atoms are those of uranium and thorium, and further, that the richest sources of radium are the uranium rocks, we conceive the possibility that radium may be a degradation-product of uranium, as emanation and helium are of radium. But until the point is definitely settled, we must keep in sight the synthetical possibilities also. As in chemistry we have more complex and less stable molecules formed out of simpler ones or out of atoms, so the radium atoms may be formed under special conditions out of smaller bodies. With so much disintegrating uranium present, special conditions are undoubtedly provided in the form of a comparatively large number of α and β particles in a state of temporary liberation from the disintegrating uranium. These may be able to recombine into large atoms, perhaps taking in as a constituent some other atoms, of some of the known elements, which happen to be present. It may be the presence of these other elements, in association with the disintegrating uranium, that determines the production of the new heavy atom of only imperfect stability, which is known to us as radium. In that case the formation of radium is not the direct result of disintegration alone, but is the final product of disintegration followed by synthesis under special conditions. Ramsay has recently given encouragement to the belief that experimental evidence may be obtained of the possibility of some action of this nature.

XVII.

Conclusion.

And now this brief account of some of the facts and problems of radium must close. The puzzles mentioned in the latter pages, as likely to meet with a solution through the further study of radium, are not a tithe of those which have presented themselves. They are enough, however, to show that it is not only for direct benefits, as in the case of lupus and rodent ulcer, that we are indebted to radium. Scientifically our great obligation to it is on the score that it has so powerfully stimulated energy, perseverance, and ingenuity in various new methods of putting nature to the question, with the result that we have already had a flood of light thrown upon atomic structure and atomic energy. And it is likely that the light will before long reach far enough to enable us to see right up to the boundaries which human knowledge is never likely to pass--the region where all kinds of energy are resolvable into the one simplest and most universal form of motion, all forms of matter into the smallest units admitting of no further analysis, and all natural laws into one simple statement of the constant relation between elementary motion and units of matter. In the practical world it is quite within the limits of reasonable hope that the early future may see us able to apply the principles learned from radium and its fellows so successfully as to effect enormous advances in the production of useful substances and economical power.

INDEX.

M'Farlane & Erskine, Printers, Edinburgh

Printed in Dunstable, United Kingdom

91220738R00080